Ben Stacy Jerrik (Ed.)

Nebraska Highway 58

AF693942

Ben Stacy Jerrik (Ed.)

Nebraska Highway 58

Loup River, U.S. Route 281, St. Paul, Nebraska, Nebraska Highway 70,

Part Press

Imprint

Publisher:
Part Press is a trademark of
International Book Market Service Ltd., 17 Rue Meldrum, Beau Bassin, 1713-01 Mauritius
Email: info@bookmarketservice.com
Website: www.bookmarketservice.com

Published in 2013

Printed in: U.S.A., U.K., Germany. This book was not produced in Mauritius.

ISBN: 978-613-9-21739-7

Contents

Articles

References

Serenus_of_Antinouplis

Serenus of Antinouplis (ca. 300 AD - ca. 360 AD) was an ancient Greek mathematician of the Roman Imperial Period.

Life and work

Serenus came either from Antinoeia or from Antinouplis, a city in Egypt founded by Hadrian. Two sources confirm that he was born in Antinouplis. It was once believed that he was born in Antissa, but this has been shown to have been based on an error.

Serenus wrote a commentary on the *Conics* of Apollonius, which is now lost. We hear from Theon of Smyrna that the main result of the commentary was that of a number of angles that are subtended at a point on a diameter of a circle that is not the center, then with equal arcs of that circle, the angle nearer to the center is always less than the angle farther away from the center.[1] But he was also a prime mathematician in his own right, having written two works entitled *On the Section of a Cylinder* and *On the Section of a Cone*, works that came to be connected to Apollonius' *Conics*. This connection helped them to survive through the ages.

In the preface of *On the Section of a Cylinder*, Serenus states that his motivation for writing this work, as summarized by Heath, was that "many persons who were students of geometry were under the erroneous that the oblique section of a cylinder was different from the oblique section of a cone known as an ellipse, whereas it is of course the same curve."[1] The work consists of thirty-three propositions.

- Proposition six proves the existence an oblique cylinder of the parallel circular sections subcontrary to the series of which the bases are two.[2]
- Proposition nine proves that the section by any plane not parallel to that of the bases or of one of the subcontrary sections but cutting all the generators is not a circle.[2]
- Propositions fourteen and sixteen, the main results, are follow ups of the previous propositions where the said section is proved to have the property of the ellipse.[2]
- Proposition seventeen puts the property that was found in propositions fourteen and sixteen into the Apollonian form using the latus rectum.[2]
- Propositions twenty-nine to thirty-three deal with an optical problem. He gives a definition of parallels that was generally ridiculed.

In propositions one though fifty-seven of *On the Section of a Cone*, Serenus deals largely with the areas of triangular sections of right and scalene cones that are created by planes passing through the vertex. He shows when the area of a triangle of a certain class of triangles is at its maximum. Propositions fifty-eight through sixty-nine form a separate section of the book and deal with the volumes of right right cones in relation to their heights, their bases and the areas of triangular sections through the axis.

Citations and footnotes

[1] Heath p. 519 (1981)
[2] Heath p. 520 (1981)

References

- Heath, Thomas Little (1981). *A History of Greek Mathematics, Volume II*. Dover publications. ISBN 0-486-24074-6.
- Ivor Bulmer-Thomas, Biography in Dictionary of Scientific Biography (New York 1970-1990).

External links

- O'Connor, John J.; Robertson, Edmund F., "Serenus of Antinouplis" (http://www-history.mcs.st-andrews.ac.uk/Biographies/Serenus.html), *MacTutor History of Mathematics archive*, University of St Andrews.

Hellenistic_civilization

The Venus de Milo, Louvre, is one of the most famous products of Hellenistic art.

The **Hellenistic period** or **Hellenisitc civilization** is the period of ancient Greek history between the death of Macedonian king Alexander the Great in 323 BC and the emergence of ancient Rome. During this time, Greek cultural influence and power was at its zenith in Europe and Asia, experiencing prosperity and progress in the arts, exploration, literature, theatre, architecture, music, mathematics, philosophy and science. It is often considered a period of transition, sometimes even of decadence or degeneration,[1] compared to the brilliance of the Greek Classical era.

After Alexander the Great's ventures in the Persian Empire, Hellenistic kingdoms were established throughout south-west Asia (Seleucid Empire, Kingdom of Pergamon) and north-east Africa (Ptolemaic Kingdom). This resulted in the export of Greek culture and language to these new realms, and moreover Greek colonists themselves. Equally, however, these new kingdoms were influenced by the indigenous cultures, adopting local practices where beneficial, necessary or convenient. Hellenistic culture thus represents a fusion of the Ancient Greek world with that of the Near East, Middle East and Southwest Asia, and a departure from earlier Greek attitudes towards "barbarian" cultures. [2] The Hellenistic periods was characterized by a new wave of Greek colonization[3] (as distinguished from that occurring in the 8th-6th centuries BC) which established Greek cities and kingdoms in Asia and Africa.[4] Those new cities were composed of Greek colonists who came from different parts of the Greek world, and not, as before, from a specific "mother city".[4] The main cultural centers expanded from mainland Greece to Pergamon, Rhodes, and new Greek colonies such as Seleucia, Antioch and Alexandria. This mixture of Greek-speakers gave birth to a common Attic-based dialect, known as Koine Greek, which became the *lingua franca* through the Hellenistic world. The Hellenistic period may be seen to end either with the final conquest of the Greek heartlands by Rome in 146 BC, with the final defeat of the Ptolemaic Kingdom at Battle of Actium in 31 BC, or even the move by Roman Emperor Constantine the Great of the capital of the Roman Empire to Constantinople in 330 AD.[5] [6] [7]

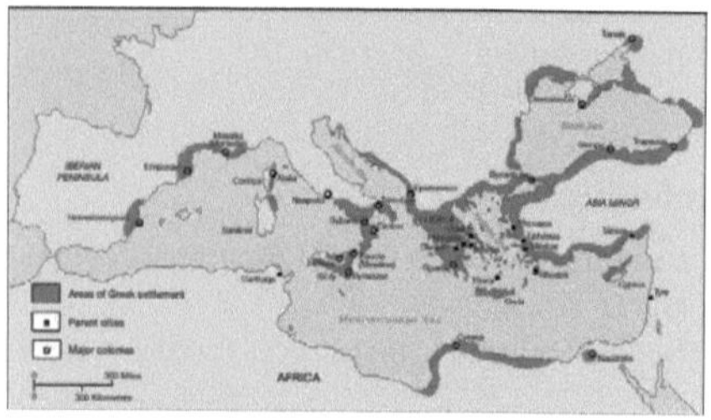

The Greek territories and colonies during the Archaic period.

Etymology

The word Hellenistic is a modern word and a 19th century concept; the idea of a Hellenistic period did not exist in Ancient Greece. In the mid-19th Century, J. G. Droysen coined the term Hellenistic to define the period when Greek culture spread in the non-Greek world after Alexander's conquest. The major issue with the term Hellenistic lies in its convenience, as the spread of Greek culture was not the generalized phenomenon that the term implies. Some areas of the conquered world were more affected by Greek influences than others. The term Hellenistic also implies that the Greek populations were of majority in the areas in which they settled, while in many cases, the Greek settlers were actually the minority amongst the native populations. The Greek population and the native population did not mix; the Greeks moved and brought their own culture, but interaction did not always occur.

Philip II, King of Macedon, gained dominance of Greece and created the League of Corinth, laying the foundation for the conquests of his son Alexander the Great.

Background

Alexander fighting the Persian king Darius III. From Alexander Mosaic, Naples National Archaeological Museum.

Ancient Greece had traditionally been a fractious collection of fiercely independent city-states. After the Peloponnesian War (431–404 BC), Greece had fallen under a Spartan hegemony, in which Sparta was pre-eminent but not all-powerful. Spartan hegemony was succeeded by a Theban one after the Battle of Leuctra (371 BC), but after the Battle of Mantinea (362 BC), all of Greece was so weakened that no one state could claim pre-eminence. It was against this backdrop, that the ascendancy of Macedon began, under king Philip II. Macedon was located at the periphery of the Greek world, and although its royal family claimed Greek descent, the Macedonians themselves were looked down upon as semi-barbaric by the rest of the Greeks. However, Macedon had a relatively strong and centralised government, and compared to most Greek states, directly controlled a large area.

With the accession of a strong and expansionist leader in Philip, Macedon was able to begin a rise to power over Greece. Philip took every opportunity to expand Macedonian territory, and in 352 BC annexed Thessaly and Magnesia. Desultory conflicts with Thebes and Athens continued for another decade but in 338 BC, Philip defeated a Theban and Athenian army at the Battle of Chaeronea. In the aftermath, Philip formed the League of Corinth, effectively bringing the majority of Greece under his direct sway. He was elected *Hegemon* of the league, and a campaign against the Achaemenid Empire of Persia was planned. However, whilst this campaign was in its early stages, he was assassinated. (possibly at the instigation of his son Alexander).[1]

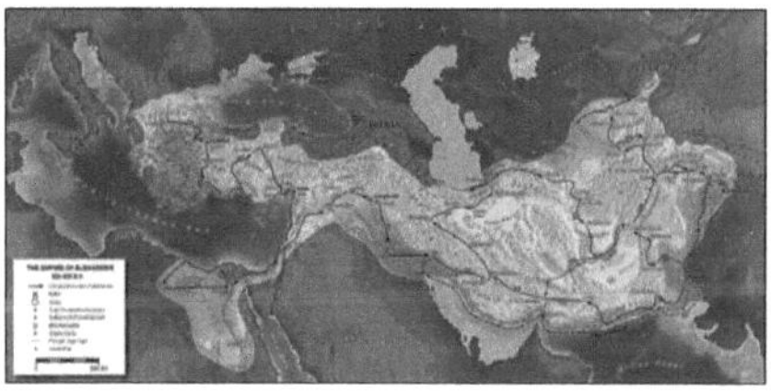

Alexander's empire at the time of its maximum expansion.

Succeeding his father, Alexander took over the Persian war himself. During a decade of campaigning, Alexander conquered the whole Persian Empire, overthrowing the Persian king Darius III. The conquered lands included Asia Minor, Assyria, the Levant, Egypt, Mesopotamia, Media, Persia, and parts of modern Afghanistan, Pakistan, and the steppes of central Asia. The years of constant campaigning had taken their toll however, and Alexander died in 323 BC.

After his death, the huge territories Alexander had conquered became subject to a strong Greek influence (hellenization) for the next two or three centuries, until the rise of Rome in the west, and of Parthia in the east. As the Greek and Levantine cultures mingled, the development of a hybrid Hellenistic culture began, and persisted even when isolated from the main centres of Greek culture (for instance, in the Greco-Bactrian kingdom).

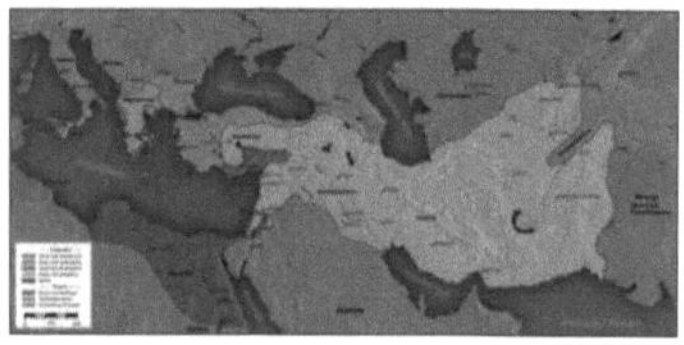

Kingdoms of the Diadochi after the battle of Ipsus, c. 301 BC. Kingdom of Ptolemy I Soter Kingdom of Cassander Kingdom of Lysimachus Kingdom of Seleucus I Nicator EpirusOther Carthage Roman Republic Greek colonies#Greek coloniesGreek colonies

It can be argued that some of the changes across the Macedonian Empire after Alexander's conquests and during the rule of the Diadochi would have occurred without the influence of Greek rule. As mentioned by Peter Green, numerous factors of conquest have been merged under the term Hellenistic Period. Specific areas conquered by Alexander's invading army, including Egypt and areas of Asia Minor and Mesopotamia "fell" willingly to conquest and viewed Alexander as more of a liberator than a victor.[8]

In addition, much of the area conquered would continue to be ruled by the Diadochi, Alexander's generals and successors. Initially the whole empire was divided amongst them; however, some territories were lost relatively quickly, or only remained nominally under Macedonian rule. After 200 years, only much reduced and rather degenerate states remained,[6] until the conquest of Ptolemaic Egypt by Rome.

The successors

When Alexander the Great died (June 10, 323 BC), he left behind a huge empire which was composed of many essentially independent territories. Alexander's empire stretched from his homeland of Macedon itself, along with the Greek city-states that his father had subdued, to Bactria and parts of India in the east. It included Anatolia, the Levant, Egypt, Babylonia, and Persia.

Without a chosen successor there was almost immediately a dispute among his generals as to who his successor should be. Meleager and the infantry supported the candidacy of Alexander's half-brother, Arrhidaeus, while Perdiccas, the leading cavalry commander, supported waiting until the birth of Alexander's unborn child by Roxana. A compromise was arranged – Arrhidaeus (as Philip III) should become king, and should rule jointly with Roxana's child, assuming that it was a boy (as it was, becoming Alexander IV). Perdiccas himself would become regent of the entire empire, and Meleager his lieutenant. Soon, however, Perdiccas had Meleager and the other infantry leaders murdered, and assumed full control.

The other cavalry generals who had supported Perdiccas were rewarded in the partition of Babylon by becoming satraps of the various parts of the empire. Ptolemy received Egypt; Laomedon received Syria and Phoenicia; Philotas took Cilicia; Peithon took Media; Antigonus received Phrygia, Lycia and Pamphylia; Asander received Caria; Menander received Lydia; Lysimachus received Thrace; Leonnatus received Hellespontine Phrygia; and Neoptolemus had Armenia. Macedon and the rest of Greece were to be under the joint rule of Antipater, who had governed them for Alexander, and Craterus, Alexander's most able lieutenant, while Alexander's old secretary, Eumenes of Cardia, was to receive Cappadocia and Paphlagonia.

In the east, Perdiccas largely left Alexander's arrangements intact – Taxiles and Porus ruled over their kingdoms in India; Alexander's father-in-law Oxyartes ruled Gandara; Sibyrtius ruled Arachosia and Gedrosia; Stasanor ruled Aria and Drangiana; Philip ruled Bactria and Sogdiana; Phrataphernes ruled Parthia and Hyrcania; Peucestas governed Persis; Tlepolemus had charge over Carmania; Atropates governed northern Media; Archon got Babylonia; and Arcesilas ruled northern Mesopotamia. By about 281 BC, the situation had stabilised, resulting in four major domains:

- The Antigonid dynasty in Macedon and central Greece;
- The Ptolemaic dynasty in Egypt based at Alexandria;
- The Seleucid dynasty in Syria and Mesopotamia based at Antioch;
- The Attalid dynasty in Anatolia based at Pergamum.

The Ptolemaic kingdom

Bust of Ptolemy I Soter, founder of the Ptolemaic Kingdom, Louvre Museum.

Ptolemy, a somatophylax, one of the seven bodyguards who served as Alexander the Great's generals and deputies, was appointed satrap of Egypt after Alexander's death in 323 BC. In 305 BC, he declared himself King Ptolemy I, later known as "Soter" (saviour). The Egyptians soon accepted the Ptolemies as the successors to the pharaohs of independent Egypt. Ptolemy's family ruled Egypt until the Roman conquest of 30 BC.

All the male rulers of the dynasty took the name Ptolemy. Ptolemaic queens, some of whom were the sisters of their husbands, were usually called Cleopatra, Arsinoe or Berenice. The most famous member of the line was the last queen, Cleopatra VII, known for her role in the Roman political battles between Julius Caesar and Pompey, and later between Octavian and Mark Antony. Her suicide at the conquest by Rome marked the end of Ptolemaic rule in Egypt.

The Seleucid Kingdom

Seleucus I Nicator, founder of the Seleucid Kingdom.

Following the carve up of Alexander's empire, Seleucus I Nicator received Babylonia From there, he created a new empire which expand to include much of Alexander's near eastern territories.[9] [10] [11] [12] At the height of its power, it included central Anatolia, the Levant, Mesopotamia, Persia, today's Turkmenistan, Pamir and parts of Pakistan.

The Seleucid Empire was a major center of Hellenistic culture which maintained the preeminence of Greek customs and where a Greek-Macedonian political elite dominated, mostly in the urban areas.[12] [13] [14] [15] The Greek population of the cities who formed the dominant elite were reinforced by emigration from Greece.[12] [13] Seleucid expansion into Anatolia and Greece was abruptly halted after decisive defeats at the hands of the Roman army. Their attempts to defeat their old enemy Ptolemaic Egypt was frustrated by Roman demands. Much of the eastern part of the empire was conquered by the Parthians under Mithridates I of Parthia in the mid-2nd century BC, yet the Seleucid kings continued to rule a rump state from Syria until the invasion by Armenian king Tigranes the Great and their ultimate overthrow by the Roman general Pompey.

Arabian Peninsula

Further information: Tylos

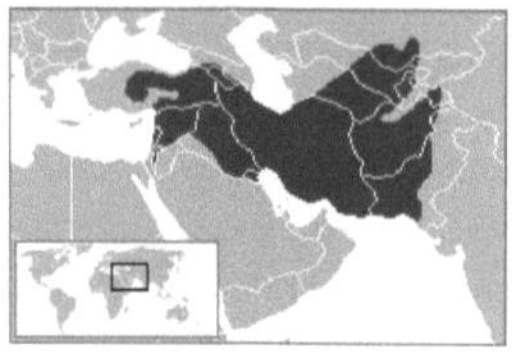

The Seleucid Empire in 300 BC.

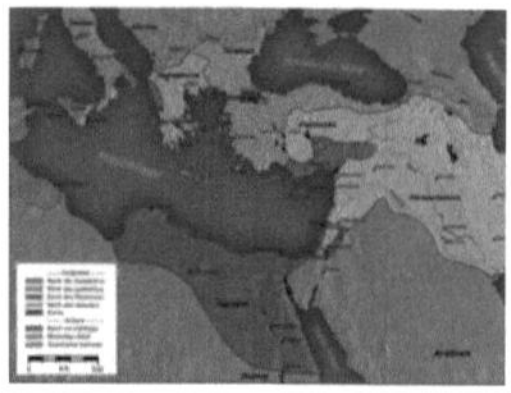

Ptolemaic Kingdom in blue, 300 BC.

Bahrain was referred to by the Greeks as **Tylos**, the centre of pearl trading, when Nearchus came to discover it serving under Alexander the Great.[16] From the 6th to 3rd century BC Bahrain was included in Persian Empire by Achaemenians, an Iranian dynasty.[17] The Greek admiral Nearchus is believed to have been the first of Alexander's commanders to visit this islands, and he found a verdant land that was part of a wide trading network; he recorded: "That in the island of Tylos, situated in the Persian Gulf, are large plantations of cotton tree, from which are manufactured clothes called *sindones*, a very different degrees of value, some being costly, others less expensive. The use of these is not confined to India, but extends to Arabia."[18] The Greek historian, Theophrastus, states that much of the islands were covered in these cotton trees and that Tylos was famous for exporting walking canes engraved with emblems that were customarily carried in Babylon.[19]

It is not known whether Bahrain was part of the Seleucid Empire, although the archaeological site at Qalat Al Bahrain has been proposed as a Seleucid base in the Persian Gulf.[20] Alexander had planned to settle the eastern shores of the Persian Gulf with Greek colonists, and although it is not clear that this happened on the scale he envisaged, Tylos was very much part of the Hellenised world: the language of the upper classes was Greek (although Aramaic was in everyday use), while Zeus was worshipped in the form of the Arabian sun-god Shams.[21] Tylos even became the site of Greek athletic contests.[22]

The name Tylos is thought to be a Hellenisation of the Semitic, Tilmun (from Dilmun).[23] The term Tylos was commonly used for the islands until Ptolemy's *Geographia* when the inhabitants are referred to as 'Thilouanoi'.[24] Some place names in Bahrain go back to the Tylos era, for instance, the residential suburb of Arad in Muharraq, is believed to originate from "Arados", the ancient Greek name for Muharraq island.[25]

The ancient Greeks speculated as to whether the Phoenicians were originally from Tylos. According to the 19th century German classicist, Arnold Heeren: "In the Greek geographers, for instance, we read of two islands, named Tyrus or Tylos, and Aradus, which boased that they were the mother country of the Phoenicians, and exhibited relics of Phoenician temples."[26] The people of Tyre in particular have long maintained Persian Gulf origins, and the similarity in the words "Tylos" and "Tyre" has been commented upon.[27]

With the waning of Seleucid Greek power, Tylos was incorporated into Characene or Mesenian, the state founded in what today is Kuwait by Hyspaosines in 127BC. A building inscriptions found in Bahrain indicate that Hyspoasines occupied the islands, (and it also mention his wife, Thalassia). From the third century BC to arrival of Islam in the seventh AD, Bahrain was controlled by two other Iranian dynasties of Parthians and Sassanids.

By about 250 BC, Seleucids lost their tritories to Parthians, an Iranian tribe from Central Asia. Parthian dynasty brought the Persian Gulf under their control and extended their influence as far as Oman. Because they needed to control the Persian Gulf trade route, the Parthians established garrisons in the southern coast of Persian Gulf.[28]

Greek peninsula

During the Hellenistic period the importance of Greece proper within the Greek-speaking world declined sharply. The great centers of Hellenistic culture were Alexandria and Antioch, capitals of Ptolemaic Egypt and Seleucid Syria respectively. Cities such as Pergamon, Ephesus, Rhodes and Seleucia were also important, and increasing urbanization of the Eastern Mediterranean was characteristic of the time.

Macedonian dominance

The quests of Alexander had a number of consequences for the Greek city-states. It greatly widened the horizons of the Greeks, making the endless conflicts between the cities which had marked the 5th and 4th centuries BC seem petty and unimportant. It led to a steady emigration, particularly of the young and ambitious, to the new Greek empires in the east. Many Greeks migrated to Alexandria, Antioch and the many other new Hellenistic cities founded in Alexander's wake, as far away as what are now Afghanistan and Pakistan, where the Greco-Bactrian Kingdom and the Indo-Greek Kingdom survived until the end of the 1st century BC.

The defeat of the Greek cities by Philip and Alexander also taught the Greeks that their city-states could never again be powers in their own right, and that the hegemony of Macedon and its successor states could not be challenged unless the city states united, or at least federated. The Greeks valued their local independence too much to consider actual unification, but they made several attempts to form federations through which they could hope to reassert their independence.

Antipatrid dynasty

Following Alexander's death a struggle for power broke out among his generals, which resulted in the break-up of his empire and the establishment of a number of new kingdoms. Macedon fell to Cassander, son of Alexander's leading general Antipater, who after several years of warfare made himself master of most of the rest of Greece. He founded a new Macedonian capital at Thessaloniki and was generally a constructive ruler.

Cassander's power was challenged by Antigonus, ruler of Anatolia, who promised the Greek cities that he would restore their freedom if they supported him. This led to successful revolts against Cassander's local rulers. In 307 BC, Antigonus's son Demetrius captured Athens and restored its democratic system, which had been suppressed by Alexander. But in 301 BC a coalition of Cassander and the other Hellenistic kings defeated Antigonus at the Battle of Ipsus, ending his challenge.

Antigonid dynasty

Hellenistic Greek tomb door bas relief, Leeds City Museum.

The Kingdom of Pergamon shown at its greatest extent in 188 BC.

After Cassander's death in 298 BC, however, Demetrius seized the Macedonian throne and gained control of most of Greece. He was defeated by a second coalition of Greek rulers in 285 BC, and mastery of Greece passed to the king Lysimachus of Thrace. Lysimachus was in turn defeated and killed in 280 BC. The Macedonian throne then passed to Demetrius's son Antigonus II, who also defeated an invasion of the Greek lands by the Gauls, who at this time were living in the Balkans. The battle against the Gauls united the Antigonids of Macedon and the Seleucids of Antioch, an alliance which was also directed against the wealthiest Hellenistic power, the Ptolemies of Egypt.

Antigonus II ruled until his death in 239 BC, and his family retained the Macedonian throne until it was abolished by the Romans in 146 BC. Their control over the Greek city states was intermittent, however, since other rulers, particularly the Ptolemies, subsidised anti-Macedonian parties in Greece to undermine the Antigonids' power. Antigonus placed a garrison at Corinth, the strategic centre of Greece, but Athens, Rhodes, Pergamum and other Greek states retained substantial independence, and formed the Aetolian League as a means of defending it. Sparta also remained independent, but generally refused to join any league.

In 267 BC, Ptolemy II persuaded the Greek cities to revolt against Antigonus, in what became the Chremonidian War, after the Athenian leader Chremonides. The cities were defeated and Athens lost her independence and her democratic institutions. The Aetolian League was restricted to the Peloponnese, but on being allowed to gain control of Thebes in 245 BC became a Macedonian ally. This marked the end of Athens as a political actor, although it remained the largest, wealthiest and most cultivated city in Greece. In 255 BC, Antigonus defeated the Egyptian fleet at Cos and brought the Aegean islands, except Rhodes, under his rule as well.

Antigonus II died in 239 BC. His death saw another revolt of the city-states of the Achaean League, whose dominant figure was Aratus of Sicyon. Antigonus's son Demetrius II died in 229 BC, leaving a child (Philip V) as king, with the general Antigonus Doson as regent. The Achaeans, while nominally subject to Ptolemy, were in effect independent, and controlled most of southern Greece. Athens remained aloof from this conflict by common consent.

Sparta remained hostile to the Achaeans, and in 227 BC Sparta's king Cleomenes III invaded Achaea and seized control of the League. Aratus preferred distant Macedon to nearby Sparta, and allied himself with Doson, who in 222 BC defeated the Spartans and annexed their city – the first time Sparta had ever been occupied by a foreign power.

Philip V, who came to power when Doson died in 221 BC, was the last Macedonian ruler with both the talent and the opportunity to unite Greece and preserve its independence against the "cloud rising in the west": the ever-increasing power of Rome. He was known as "the darling of Hellas". Under his auspices the Peace of Naupactus (217 BC) brought conflict between Macedon and the Greek leagues to an end, and at this time he controlled all of Greece except Athens, Rhodes and Pergamum.

In 215 BC, however, Philip formed an alliance with Rome's enemy Carthage, which drew Rome directly into Greek affairs for the first time. Rome promptly lured the Achaean cities away from their nominal loyalty to Philip, and formed alliances with Rhodes and Pergamum, now the strongest power in Asia Minor. The First Macedonian War broke out in 212 BC, and ended inconclusively in 205 BC, but Macedon was now marked as an enemy of Rome. Rome's ally Rhodes gained control of the Aegean islands.

In 202 BC, Rome defeated Carthage, and was free to turn her attention eastwards, urged on by her Greek allies, Rhodes and Pergamum. In 198 BC, the Second Macedonian War broke out for obscure reasons, but very likely because Rome saw Macedon as a potential ally of the Seleucids, the greatest power in the east. Philip's allies in Greece deserted him and in 197 BC he was decisively defeated at the Cynoscephalae by the Roman proconsul Titus Quinctius Flamininus.

Luckily for the Greeks, Flamininus was a moderate man and an admirer of Greek culture. Philip had to surrender his fleet and become a Roman ally, but was otherwise spared. At the Isthmian Games in 196 BC, Flamininus declared all the Greek cities free, although Roman garrisons were placed at Corinth and Chalcis. But the freedom promised by Rome was an illusion. All the cities except Rhodes were enrolled in a new League which Rome ultimately controlled, and democracies were replaced by aristocratic regimes allied to Rome.

Roman expansion

In 192 BC, war broke out between Rome and the Seleucid ruler Antiochus III. Antiochus invaded Greece with a 10,000 man army, and was elected the commander in chief of the Aetolians . Some Greek cities now thought of Antiochus as their saviour from Roman rule, but Macedon threw its lot in with Rome. In 191 BC, the Romans under Manius Acilius Glabrio routed him at Thermopylae and obliged him to withdraw to Asia. During the course of this war Roman troops moved into Asia for the first time, where they defeated Antiochus again at Magnesia on the Sipylum (190 BC). Greece now lay across Rome's line of communications with the east, and Roman soldiers became a permanent presence. The Peace of Apamaea (188 BC) left Rome in a dominant position throughout Greece.

Bust of Antiochus III the Great from the Louvre.

During the following years Rome was drawn deeper into Greek politics, since the defeated party in any dispute appealed to Rome for help. Macedon was still independent, though nominally a Roman ally. When Philip V died in 179 BC, he was succeeded by his son Perseus, who like all the Macedonian kings dreamed of uniting the Greeks under Macedonian rule. Macedon was now too weak to achieve this objective, but Rome's ally Eumenes II of Pergamum persuaded Rome that Perseus was a potential threat to Rome's position.

End of Greek independence

As a result of Eumenes's intrigues Rome declared war on Macedon in 171 BC, bringing 100,000 troops into Greece. Macedon was no match for this army, and Perseus was unable to rally the other Greek states to his aid. Poor generalship by the Romans enabled him to hold out for three years, but in 168 BC the Romans sent Lucius Aemilius Paullus to Greece, and at Pydna the Macedonians were crushingly defeated. Perseus was captured and taken to Rome, the Macedonian kingdom was broken up into four smaller states, and all the Greek cities who aided her, even rhetorically, were punished. Even Rome's allies Rhodes and Pergamum effectively lost their independence.

Under the leadership of an adventurer called Andriscus, Macedon rebelled against Roman rule in 149 BC: as a result it was directly annexed the following year and became a Roman province, the first of the Greek states to suffer this fate. Rome now demanded that the Achaean League, the last stronghold of Greek independence, be dissolved. The Achaeans refused and, feeling that they might as well die fighting, declared war on Rome. Most of the Greek cities rallied to the Achaeans' side, even slaves were freed to fight for Greek independence. The Roman consul Lucius Mummius advanced from Macedonia and defeated the Greeks at Corinth, which was razed to the ground.

In 146 BC, the Greek peninsula, though not the islands, became a Roman protectorate. Roman taxes were imposed, except in Athens and Sparta, and all the cities had to accept rule by Rome's local allies. In 133 BC, the last king of Pergamum died and left his kingdom to Rome: this brought most of the Aegean peninsula under direct Roman rule as part of the province of Asia.

The final downfall of Greece came in 88 BC, when King Mithridates of Pontus rebelled against Rome, and massacred up to 100,000 Romans and Roman allies across Asia Minor. Although Mithridates was not Greek, many Greek cities, including Athens, overthrew their Roman puppet rulers and joined him. When he was driven out of Greece by the Roman general Lucius Cornelius Sulla, Roman vengeance fell upon Greece again, and the Greek cities never recovered. Mithridates was finally defeated by Gnaeus Pompeius Magnus (Pompey the Great) in 65 BC .

Further ruin was brought to Greece by the Roman civil wars, which were partly fought in Greece. Finally, in 27 BC, Augustus directly annexed Greece to the new Roman Empire as the province of Achaea. The struggles with Rome had left Greece depopulated and demoralised. Nevertheless, Roman rule at least brought an end to warfare, and cities such as Athens, Corinth, Thessaloniki and Patras soon recovered their prosperity.

The Greco-Bactrians

The Greco-Bactrian kingdom at its maximum extent (c. 180 BC)

The Greek kingdom of Bactria (or Greco-Bactrian kingdom) began as an offshoot of the Seleucid empire. The sheer size of the eastern Seleucid domains must mean that the satraps governing the provinces had significant freedom from central control. In around 250 BC, the governor of Bactria, Sogdiana and Margiana, one Diodotus, took this process to its logical extreme and declared himself king. At around the same time, the re-emergence of a native Persian dynasty under the Parthian king Arsaces effectively cut the nascent Greco-Bactrian kingdom off from the rest of the Seleucid empire. This probably allowed it to maintain its independence in the medium term, but in the long-term may have contributed to its decline and fall; it could no longer receive manpower or aid from other Hellenistic regions at sufficient levels.

Diodotus II, son of Diodotus, was overthrown in about 230 BC by Euthydemus, possibly the satrap of Sogdiana, who then started his own dynasty. In c. 210 BC, the Greco-Bactrian kingdom was invaded by a resurgent Seleucid empire under Antiochus III. Whilst victorious in the field, it seems Antiochus came to realise that there were advantages in the status quo (perhaps sensing that Bactria could not be governed from Syria), and married one of his daughters to Euthydemus's son, thus legitimising Greco-Bactria. Soon afterwards the Greco-Bactrian kingdom seems to have expanded, possibly taking advantage of the defeat of the Parthian king Arsaces II by Antiochus.

Coin of Demetrius the Invincible, Metropolitan Museum of Art.

Demetrius, son and successor of Euthydemus, invaded north-western India in 180 BC, after the destruction of the Mauryan empire there; the Mauryans were probably allies of the Bactrians (and Seleucids). The exact justification for the invasion remains unclear, but by about 175 BC, the Greeks ruled over parts of north-western India.

This period also marks the beginning of the obfuscation of Greco-Bactrian history. Demetrius possibly died about 180 BC; numismatic evidence suggest the existence of several other kings shortly thereafter. It is probable that at this point that the Greco-Bactrian kingdom split into several semi-independent regions for some years; Euthydemus II (son of Demetrius?) seems to have ruled in Bactria, with Agathocles, Antimachus I and Pantaleon ruling in India. In around 171 BC the usurper Eucratides I swept to power in Bactria, removing whichever king(s) were actually ruling at that point. Similarly, in India, the general Apollodotus I seems to have assumed more-or-less complete power by around 170 BC, thereby marking the true start of the Indo-Greek kingdom (see below).

Eucratides may have been a member of the Seleucid royal family, who set out to (re)claim the Bactrian lands. Eucratides certainly had a vast and prestigious coinage, suggesting he was a ruler of considerable importance. He appears to have re-invigorated the Bactrian kingdom, although territory was lost to Parthia in the west. He fought with the Indo-Greeks, and appears to have occupied India up to the river Indus for a while. However, his murder in 145 BC triggered a civil war which fatally weakened the kingdom as his sons Eucratides II and Heliocles I fought each other. Heliocles was the last Greek to clearly rule Bactria, his power collapsing in the face of tribal invasions of Bactria, by about 130 BC. However, Greek urban civilisation seems to have continued in Bactria after the fall of the kingdom, having a hellenising effect on the tribes which had displaced Greek-rule.

The Indo-Greek kingdoms

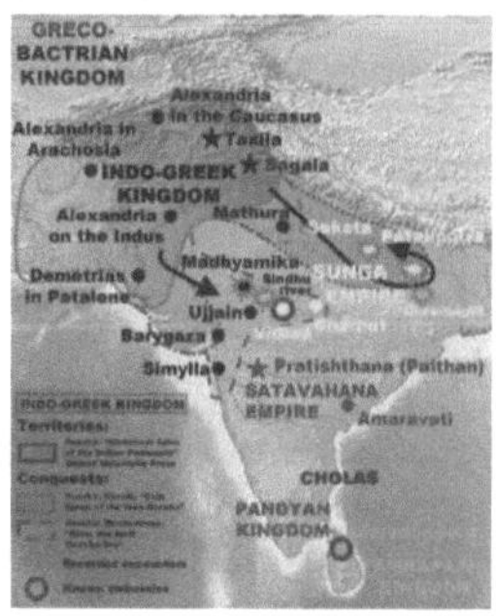

Indo-Greek territory, with known campaigns and battles.[29] [30] [31]

The separation of the Indo-Greek kingdom from the Greco-Bactrian kingdom resulted in an even more isolated position, and thus the details of the Indo-Greek kingdom are even more obscure than for Bactria. Many supposed kings in India are known only because of coins bearing their name. The numismatic evidence together with archaeological finds and the scant historical records suggest that the fusion of eastern and western cultures reached its peak in the Indo-Greek kingdom.

As mentioned, Apollodotus made himself king of 'India' in around 170 BC. The exact fate of Apollodotus is unknown, but he seems to have extended the conquests east into Gandhara and western Punjab. In about 155 (or 165) BC he seems to have been succeeded by the most successful of the Indo-Greek kings, Menander I. Menander converted to Buddhism, and seems to have been a great patron of the religion; he is remembered in some Buddhist texts as 'Milinda'. He also expanded the kingdom further east into Punjab, though these conquests were rather ephemeral.

After the death of Menander (c. 130 BC), the Kingdom appears to have fragmented, with several 'kings' attested contemporaneously in different regions. This inevitably weakened the Greek position, and territory seems to have been lost progressively. Around 70 BC, the western regions of Arachosia and Paropamisadae were lost to tribal invasions, presumably by those tribes responsible for the end of the Bactrian kingdom. The resulting Indo-Scythian kingdom seems to have gradually pushed the remaining Indo-Greek kingdom towards the east. The Indo-Greek kingdom appears to have lingered on in western Punjab until about 10 AD when finally ended by the Indo-Scythians.

For centuries after the demise of the Indo-Greek kingdom, Greek civilization influenced Indian culture, art, and language. Even the depictions of Buddha appear to have been influenced by Greek culture: Buddha representations in the Ghandara period were often showing Buddha under the protection of Herakles.[32]

Influence of Indian culture

Several references in Indian literature praise the knowledge of the Yavanas or the Greeks. The Mahabharata compliments them as "the all-knowing Yavanas" (sarvajnaa yavanaa) i.e. "The Yavanas, O king, are all-knowing; the Suras are particularly so. The mlecchas are wedded to the creations of their own fancy."[33] and the creators of flying machines that are generally called vimanas.[12]

The "Brihat-Samhita" of the mathematician Varahamihira says: "The Greeks, though impure, must be honored since they were trained in sciences and therein, excelled others....." .[34]

Yet another Indian text, (Gargi-Samhita), also similarly compliments the Yavanas saying: "The Yavanas are barbarians yet the science of astronomy originated with them and for this they must be revered like gods".[35]

The Kingdom of Pontus

The Kingdom of Pontus was a Hellenistic kingdom on the southern coast of the Black Sea. It was founded by Mithridates I in 291 BC and lasted until its conquest by the Roman Republic in 63 BC. Despite being ruled by a dynasty which was a descendant of the Persian Achaemenid Empire it became hellenized due to the influence of the Greek cities on the Black Sea and its neighboring kingdoms. The kingdom grew to its largest extent under Mithridates VI the great, who conquered Colchis, Cappadocia, Bithynia, Lesser Armenia, the Greek colonies of the Tauric Chersonesos and for a brief time the Roman province of Asia. After a long struggle with Rome in the Mithridatic wars, Pontus was defeated, part of it was incorporated into the Roman Republic as the province Bithynia and Pontus and the eastern half survived as a client kingdom.

The Kingdom of Pontus at its height: before the reign of Mithridates VI (dark purple), after his early conquests (purple), and his conquests in the first Mithridatic wars (pink)

Rise of Rome

Widespread Roman interference in the Greek world was probably inevitable given the general manner of the ascendency of the Roman Republic. This Roman-Greek interaction began as a consequence of the Greek city-states located along the coast of southern Italy. Rome had come to dominate the Italian peninsula, and desired the submission of the Greek cities to its rule. Although they initially resisted, allying themselves with Pyrrhus of Epirus, and defeating the Romans at several battles, the Greek cities were unable to maintain this position and were absorbed by the Roman republic. Shortly afterwards, Rome became involved in Sicily, fighting against the Carthaginans in the First Punic War. The end result was the complete conquest of Sicily, including its previously powerful Greek cities, by the Romans.

The independent cities of Magna Graecia did not form part of the Hellenistic domains and had, by this time, been eclipsed in power by the Hellenistic kingdoms of the east. They also remained independent at a time when the Mediterranean was increasingly dominated by 'great powers'. This, and their proximity to Rome, had made them easy and obvious targets. Conversely, the major Hellenistic realms were not in the immediate Roman sphere of influence, and were powerful enough to deter Roman aggression. The events which, in retrospect, marked the beginning of the end for the Hellenistic kingdoms could have been avoided; even if it seems likely that a collision between them and Rome would have ultimately occurred.

Roman entanglement in the Balkans began, as so often, with trade. Illyrian piratical raids on Roman merchants twice led to a Roman task force invading Illyria (the First and, Second Illyrian Wars). Tension between Macedon and Rome increased when the young king of Macedon, Philip V harboured one of the chief pirates, Demetrius of Pharos[36] (a former client of Rome). As a result, in an attempt to reduce Roman influence in the Balkans, Philip allied himself with Carthage after Hannibal had dealt the Romans a massive defeat at the Battle of Cannae (216 BC) during the Second Punic War. Forcing the Romans to fight on another front when they were at a nadir of manpower gained Philip the lasting enmity of the Romans; the only real result from the somewhat insubstantial First Macedonian War (215–202 BC).

Once the Second Punic War had been resolved, and the Romans had begun to regather their strength, they looked to re-assert their influence in the Balkans, and to curb the expansion of Philip. A pretext for war was provided by Philip's refusal to end his war with Attalid Pergamum, and Rhodes, both Roman allies.[37] The Romans, also allied with the Aetolian League of Greek city-states (which resented Philip's power), thus declared war on Macedon in 200 BC, starting the Second Macedonian War. This ended with a decisive Roman victory at the Battle of Cynoscephalae (197 BC). Like most Roman peace treaties of the period, the resultant 'Peace of Flaminius' was designed utterly to crush the power of the defeated party; a massive indemnity was levied, Philip's fleet was surrendered to Rome, and Macedon was effectively returned to its ancient boundaries, losing influence over the city-states of southern Greece, and land in Thrace and Asia Minor. The result was the end of Macedon as a major power in the Mediterranean.

As a result of the confusion in Greece at the end of the Second Macedonian War, the Seleucid Empire also became entangled with the Romans. The Seleucid Antiochus III had allied with Philip V of Macedon in 203 BC, agreeing that they should jointly conquer the lands of the boy-king of Egypt, Ptolemy V. After defeating Ptolemy in the Fifth Syrian War, Antiochus concentrated on occupying the Ptolemaic possessions in Asia Minor. However, this brought Antiochus into conflict with Rhodes and Pergamum, two important Roman allies, and began a 'cold war' between Rome and Antiochus (not helped by the presence of Hannibal at the Seleucid court).[1] Meanwhile, in mainland Greece, the Aetolian League, which had sided with Rome against Macedon, now grew to resent the Roman presence in Greece. This presented Antiochus III with a pretext to invade Greece and 'liberate' it from Roman influence, thus starting the Roman-Syrian War (192–188 BC). Another decisive Roman victory at the Battle of Magnesia (190 BC) saw the defeat of Antiochus. Another crippling treaty followed, with Seleucid possessions in Asia Minor removed and given to Rhodes and Pergamum, the size of the Seleucid navy reduced, and a massive war indemnity invoked.

Thus, in less than twenty years, Rome had destroyed the power of one of the successor states, crippled another, and firmly entrenched its influence over Greece. This was primarily a result of the over-ambition of the Macedonian kings, and their unintended provocation of Rome; though Rome was quick to exploit the situation. In another twenty years, the Macedonian kingdom was no more. Seeking to re-assert Macedonian power and Greek independence, Philip V's son Perseus incurred the wrath of the Romans, resulting in the Third Macedonian War (171-168 BC). Victorious, the Romans abolished the Macedonian kingdom, replacing it with four puppet republics; these lasted a further twenty years before Macedon was formally annexed as a Roman province (146 BC).

The Attalid dynasty of Pergamum lasted little longer; a Roman ally until the end, its final king Attalus III died in 133 BC without an heir, and taking the alliance to its natural conclusion, willed Pergamum to the Roman Republic.[38]

Contrarily, having so firmly intricated themselves into Greek affairs, the Romans now completely ignored the rapidly disintegrating Seleucid empire (perhaps because it posed no threat); and left the Ptolemaic kingdom to decline quietly, whilst acting as a protector of sorts, in as much as to stop other powers taking Egypt over (including the famous line-in-the-sand incident when the Seleucid Antiochus IV Epiphanes tried to invade Egypt).[1] Eventually, instability in the near east resulting from the power vacuum left by the collapse of the Seleucid empire caused the Roman proconsul Pompey the Great to abolish the Seleucid rump state, absorbing much of Syria into the

Roman republic.[38] Famously, the end of Ptolemaic Egypt came as the final act in the republican civil war between the Roman triumvirs Mark Anthony and Augustus Caesar. After the defeat of Anthony and his lover, the last Ptolemaic monarch, Cleopatra VII at the Battle of Actium, Augustus invaded Egypt and took it as his own personal fiefdom.[38] He thereby completed both the destruction of the Hellenistic kingdoms and the Roman republic, and ended (in hindsight) the Hellenistic era.

Culture

In some fields Hellenistic culture thrived, particularly in its preservation of the past. As has been noted, the states of the Hellenistic period were deeply fixated with the past and its seemingly lost glories.[40]

The Library of Alexandria in the Ptolemaic Kingdom was the largest and most significant library of the ancient world, and was regarded as one of the Seven Wonders of the Ancient World.[39]

Athens retained its position as the most prestigious seat of higher education, especially in the domains of philosophy and rhetoric, with considerable libraries.[41] Alexandria was arguably the second most important center of Greek learning. The Library of Alexandria had 700,000 volumes.[41] The city of Pergamon became a major center of book production, possessing a library of some 200,000 volumes, second only to Alexandria's.[41] The island of Rhodes boasted a famous finishing school for politics and diplomacy. Cicero was educated in Athens and Mark Antony in Rhodes.[41] Antioch was founded as a metropolis and center of Greek learning which retained its status into the era of Christianity.[41] Seleucia replaced Babylon as the metropolis of the lower Tigris.

The Lighthouse in Ptolemaic Alexandria was one of the tallest man-made structures on Earth for many centuries, and was regarded as one of the Seven Wonders of the Ancient World.

The spread of Greek culture throughout the Near East and Asia owed much to the development of cities. Settlements such as Ai-Khanoum, situated on trade routes, allowed cultures to mix and spread. The identification of local gods with similar Greek deities facilitated the building of Greek-style temples, and the Greek culture in the cities also meant that buildings such as gymnasia became common. Many cities maintained their autonomy while under the nominal rule of the local king or satrap, and often had Greek-style institutions. Greek dedications, statues, architecture and inscriptions have all been found. However, local cultures were not replaced, and often mixed to create a new culture.

Greek language and literature spread throughout the former Persian Empire. The development of the Alexander Romance (mainly in Egypt) owes much to Greek theater as well as other styles of story. The Library at Alexandria, set up by Ptolemy I Soter, became a center for learning and was copied by various other monarchs. An example that shows the spread of Greek theater is Plutarch's story of the death of Crassus, in which his head was taken to the Parthian court and used as a prop in a performance of The Bacchae. Theaters have also been found: for example, in Ai-Khanoum on the edge of Bactria, the theater has 35 rows - larger than the theater in Babylon.

The spread of Greek influence and language is also shown through Ancient Greek coinage. Portraits became more realistic, and the obverse of the coin was often used to display a propaganda image, commemorating an event or displaying the image of a favored god. The use of Greek-style portraits and Greek language continued into the Parthian period, even as the use of Greek was in decline.

The Colossus of Rhodes, depicting the Greek Titan Helios, is considered one of the Seven Wonders of the Ancient World.

Hellenization

Further information: Hellenization

The concept of Hellenization, meaning the spread of Greek culture, has long been controversial. Undoubtedly Greek influence did spread through the Hellenistic realms, but to what extent, and whether this was a deliberate policy or mere cultural diffusion, have been hotly debated.

The Rosetta Stone, which provided the key to the modern understanding of Egyptian hieroglyphs, British Museum.

It seems likely that Alexander himself pursued deliberate 'Hellenization' policies. Whilst it may have been a deliberate attempt to spread Greek culture, it is more likely that it was a series of pragmatic measures designed to aid in the rule of his enormous empire.[42] These policies can also be interpreted as the result of Alexander's probable megalomania[43] during his later years.

The first tenet of Alexander's policies was the founding (or re-founding) of cities across the empire. This has, in the past, been interpreted as part of Alexander's desire to spread Greek culture throughout the empire. These cities were presumably intended to be administrative headquarters in the regions, and to have been settled by Greeks; many were settled by veterans of Alexander's campaigns. Undoubtedly, this would have resulted in the spread of Greek influence across the empire; however, the primary purpose could have been to control his new subjects, rather than specifically to spread Greek culture. Arrian explicitly says that a city founded in Bactria was "meant to civilise the natives"; however, this comment could be interpreted in either way (with civilise as a euphemism for 'control'). Certainly, the cities would have been garrison points, and thus allowed control of the surrounding areas.

Secondly, Alexander attempted to create a unified ruling class of Persians and Greeks, bound by marriage ties. He used both Greeks and Persians in positions of power, although he depended more on Greeks in unstable positions, and also replaced many Persian satraps in a purge after his return from India. He also tried to mix the two cultures, adopting elements of the Persian court (such as a version of the royal robes and some of the court ceremony and attendants) and also attempting to insist on the practice of proskynesis for his Greek subjects. This is probably an attempt to equalize the two races in their behavior towards Alexander as 'Great King', but it was bitterly resented by the Macedonians, as the Greek custom was reserved solely for the gods. This policy can be interpreted as an attempt to spread Greek culture, or to create a hybrid culture. However, again, it is probably better seen as an attempt to help control the unwieldy empire;[42] Alexander needed loyalty from Persian nobles as much as from his Macedonian officers. A hybrid court culture may have been created so as not to exclude the Persians. Furthermore, Alexander's marriage to, and child with the Bactrian princess Roxana can be interpreted as an attempt to create a royal dynasty which would be acceptable to both Asians and Greeks.

Alexander also unified the army, placing Persian soldiers (some trained in the Macedonian way of fighting and some in their original styles) in the Macedonian ranks. However, again, this can simply be seen as a pragmatic solution to chronic manpower problems. Alexander's increasing megalomania can be seen in his plan to completely homogenize the populations of Europe and Asia by mass re-settlement.[43] Whilst this thoroughly impractical plan could be interpreted as an attempt to create a new hybrid culture, the sheer ambitiousness of the plan suggests some other process at work.

In short, Alexander's policies did undoubtedly result in the spread of Greek culture, but whether this was their primary aim must remain doubtful. They probably represent, instead, pragmatic attempts by Alexander to control his extensive new territories, in part by presenting himself as the heir to both Greek and Asian legacies, rather than an outsider.

After Alexander's death in 323BC, the Empire was split into satrapies under his generals. Most of Alexander's cultural changes were rejected by the Diadochi, including the cross-cultural marriages they entered into.[42] However, the influx of Greek colonists into the new realms continued to spread Greek culture into Asia. The founding of new cities continued to be a major part of the Successors' struggle for control of any particular region, and these continued to be centres of cultural diffusion. The spread of Greek culture under the Successors seems mostly to have occurred with the spreading of Greeks themselves, rather than as an active policy.

Despite their initial reluctance, the Successors seem to have later deliberately naturalised themselves to their different regions, presumably in order to help maintain control of the population.[44] Thus, for instance, we find the Ptolemies, as early as Ptolemy I Soter, the first Hellenistic king of Egypt, portrayed as pharaohs. Similarly, in the Indo-Greek kingdom, we find kings who were converts to Buddhism (e.g. Menander). The Greeks in the regions therefore gradually become 'localised', adopting local customs as appropriate. In this way, hybrid 'Hellenistic' cultures naturally emerged, at least amongst the upper echelons of society.

In summary, Alexander's conquests and the Successor kingdoms allowed widespread Greek colonisation and cultural diffusion, but it is unlikely this was ever a deliberate policy. Furthermore, such Hellenization was accompanied by the opposite spread of Asian culture to Europe. Nevertheless, the upheavals which occurred during this period do seem to have resulted in the development of hybrid 'Hellenistic' cultures. As a final point, it should be noted that the degree of influence that Greek culture had throughout the Hellenistic regions is often exaggerated because of the great influence on later generations of a small number of extensively Hellenized cities, particularly Alexandria.

Religion

In the Hellenistic period, there was much continuity in Greek religion: the Greek gods continued to be worshiped, and the same rites were practiced as before. Change came from the addition of new religions from other countries, such as including the Egyptian god(esse)s of Isis and Serapis, and the Syrian gods of Atargatis and of Hadad, which provided a new outlet for people seeking fulfillment in both the present life and the afterlife. The worship of Hellenistic rulers was also a feature of this period, most notably in Egypt, where the Ptolemies adopted earlier pharaonic practice, and established themselves as god-kings. Elsewhere rulers might receive divine status without the full status of a god.

Remains of the temple of Apollo at Corinth.

Magic was practiced widely, and these too, were a continuation from earlier times. Throughout the Hellenistic world, people would consult oracles, and use charms and figurines to deter misfortune or to cast spells. Also developed in this era was the complex system of astrology, which sought to determine a person's character and future in the movements of the sun, moon, and planets. The systems of Hellenistic philosophy, such as Stoicism and Epicureanism, offered an alternative to traditional religion, even if their impact was largely limited to the educated elite.

Philosophy

During the Hellenistic and Roman periods, many different schools of thought developed in the Hellenistic world and then the Greco-Roman world. Hellenistic philosophy experienced an age of eclecticism, a new awakening of the diverse knowledge and theories present in Greek culture. Instead of contemplating and debating ideals, logic, extinguished emotion, or consummate beauty, people would explore and analyze reality.[5] There were Greeks, Romans, Egyptians, Syrians and Arabs who contributed to the development of Hellenistic philosophy. Elements of Persian philosophy and Indian philosophy also had an influence. The most notable schools of Hellenistic philosophy were:

Zeno of Citium founded Stoic philosophy.

- Neoplatonism: Plotinus (Egyptian), Ammonius Saccas, Porphyry (Syrian), Zethos (Arab), Iamblichus (Syrian), Proclus
- Academic Skepticism: Arcesilaus, Carneades, Cicero (Roman)
- Pyrrhonian Skepticism: (?) Pyrrho, Sextus Empiricus
- Cynicism: Antisthenes, Diogenes of Sinope, Crates of Thebes (taught Zeno of Citium, founder of Stoicism)
- Stoicism: Zeno of Citium, Cleanthes, Chrysippus, Crates of Mallus (brought Stoicism to Rome c. 170 BCE), Panaetius, Posidonius, Seneca (Roman), Epictetus (Greek/Roman), Marcus Aurelius (Roman)
- Epicureanism: Epicurus (Greek) and Lucretius (Roman)
- Eclecticism: (?) Cicero (Roman)

The spread of Christianity throughout the Roman world, followed by the spread of Islam, ushered in the end of Hellenistic philosophy and the beginnings of Medieval philosophy, which was dominated by the three Abrahamic traditions: Jewish philosophy, Christian philosophy, and early Islamic philosophy.

Sciences

Further information: Hellenistic astronomy, Hellenistic mathematics, and Hellenistic geography

Archimedes Thoughtful by Fetti (1620).

Hellenistic culture produced seats of learning in Alexandria in Egypt and Antioch in Syria along with Greek speaking populations across several monarchies. Hellenistic science differed from Greek science in at least two ways: first, it benefited from the cross-fertilization of Greek ideas with those that had developed in the larger Hellenistic world; secondly, to some extent, it was supported by royal patrons in the kingdoms founded by Alexander's successors. Especially important to Hellenistic science was the city of Alexandria in Egypt, which became a major center of scientific research in the 3rd century BC. Two institutions established there during the reigns of Ptolemy I Soter (reigned 323 - 283 BC) and Ptolemy II Philadelphus (reigned 281 - 246 BC) were the Library and the Museum. Unlike Plato's Academy and Aristotle's Lyceum, these institutions were officially supported by the Ptolemies; although the extent of patronage could be precarious, depending on the policies of the current ruler.[46]

Hero of Alexandria is considered the greatest experimenter of antiquity.[45]

Hellenistic scholars frequently employed the principles developed in earlier Greek thought: the application of mathematics and deliberate empirical research, in their scientific investigations.[47]

In medicine, Herophilos (335 - 280 BC) was the first to base his conclusions on dissection of the human body and to describe the nervous system.

Geometers such as Archimedes (ca. 287 BC – 212 BC), Apollonius of Perga (ca. 262 BC – ca. 190 BC), and Euclid (ca. 325 BC – 265 BC), whose *Elements* became the most important textbook in mathematics until the 19th century, built upon the work of the Hellenic era Pythagoreans. Eratosthenes used his knowledge of geometry to measure the distance between the Sun and the Earth along with the size of the Earth.

Astronomers like Hipparchus (ca. 190 – ca. 120 BC) built upon the measurements of the Babylonian astronomers before him, to measure the precession of the Earth. Pliny reports that Hipparchus produced the first systematic star catalog after he observed a new star (it is uncertain whether this was a nova or a comet) and wished to preserve astronomical record of the stars, so that other new stars could be discovered.[48] It has recently been claimed that a celestial globe based on Hipparchus's star catalog sits atop the broad shoulders of a large 2nd-century Roman statue known as the Farnese Atlas.[49]

The level of Hellenistic achievement in astronomy and engineering is impressively shown by the Antikythera mechanism (150-100 BC). It is a 37-gear mechanical computer which computed the motions of the Sun and Moon, including lunar and solar eclipses predicted on the basis of astronomical periods believed to have been learned from the Babylonians.[50] Devices of this sort are not found again until the 10th century, when a simpler eight-geared luni-solar calculator incorporated into an astrolabe was described by the Persian scholar, Al-Biruni.[51] Similarly complex devices were also developed by other Muslim engineers and astronomers during the Middle Ages.[50]

The interpretation of Hellenistic science varies widely. At one extreme is the view of the English classical scholar, Cornford, who believed that "all the most important and original work was done in the three centuries from 600 to 300 BC"[52] At the other is the view of the Italian physicist and mathematician, Lucio Russo, who claims that scientific method was actually born in the 3rd century BC, to be forgotten during the Roman period and only revived in the Renaissance.[53]

Art

A number of the best-known works of Greek sculpture belong to the Hellenistic period, including Laocoön and his Sons, Venus de Milo, and the Winged Victory of Samothrace.

Laocoön and His Sons, Vatican Museums, Rome.

The term *Hellenistic* is a modern invention; the Hellenistic World not only included a huge area covering the whole of the Aegean, rather than the Classical Greece focused on the Poleis of Athens and Sparta, but also a huge time range. In artistic terms this means that there is huge variety which is often put under the heading of "Hellenistic Art" for convenience.

There has been a trend in writing the history of this period to depict Hellenistic art as a decadent style, following of the *Golden Age* of Classical Athens. Pliny the Elder, after having described the sculpture of the classical period says: *Cessavit deinde ars* ("then art disappeared").[54] The 18th century terms Baroque and Rococo have sometimes been applied, to the art of this complex and individual period. The renewal of the historiographic approach as well as some recent discoveries, such as the tombs of Vergina, allow a better appreciation of this period's artistic richness.

The Hellenistic period and modern culture

Further information: Decadence and Degeneration

The focus on the Hellenistic period over the course of the 19th century by scholars and historians has led to an issue common to the study of historical periods; historians see the period of focus as a mirror of the period in which they are living. Many 19th century scholars contended that the Hellenistic period represented a cultural decline from the brilliance of classical Greece. Though this comparison is now seen as unfair and meaningless, it has been noted that even commentators of the time saw the end of a cultural era which could not be matched again.[55] This may be inextricably linked with the nature of government. It has been noted by Herodotus that after the establishment of the Athenian democracy:

> *...the Athenians found themselves suddenly a great power. Not just in one field, but in everything they set their minds to...As subjects of a tyrant, what had they accomplished?...Held down like slaves they had shirked and slacked; once they had won their freedom, not a citizen but he could feel like he was labouring for himself*"[56]

Thus, with the decline of the Greek polis, and the establishment of monarchical states, the environment and social freedom in which to excel may have been reduced.[57] A parallel can be drawn with the productivity of the city states of Italy during the Renaissance, and their subsequent decline under autocratic rulers.

However, William Woodthorpe Tarn, between World War I and World War II and the heyday of the League of Nations, focused on the issues of racial and cultural confrontation and the nature of colonial rule. Michael Rostovtzeff, who fled the Russian Revolution, concentrated predominantly on the rise of the capitalist bourgeoisie in areas of Greek rule. Arnaldo Monigliano, an Italian Jew who wrote before and after the Second World War, studied the problem of mutual understanding between races in the conquered areas. Moses Hadas portrayed an optimistic picture of synthesis of culture from the perspective of the 1950s, while Frank William Walbank in the 1960s and 1970s had a materialistic approach to the Hellenistic period, focusing mainly on class relations. Recently, however, papyrologist C. Préaux has concentrated predominantly on the economic system, interactions between kings and cities and provides a generally pessimistic view on the period. Peter Green, on the other hand, writes from the point of view of late 20th century liberalism, his focus being on individualism, the breakdown of convention, experiments and a postmodern disillusionment with all institutions and political processes.[8]

See also

- Roman Republic
- Parthian Empire
- Maurya Empire
- Han dynasty
- Pre-Roman Iron Age
- Ancient Carthage
- La Tène culture
- Greco-Roman world

References

[1] Alexander The Great and the Hellenistic Age. Green P. ISBN 978-0-7538-2413-9

[2] Green, p. xvii.

[3] Professor Gerhard Rempel, *Hellenistic Civilization* (Western New England College) (http://mars.wnec.edu/~grempel/courses/wc1/lectures/10hellenism.html).

[4] Ulrich Wilcken, *Griechische Geschichte im Rahmen der Alterumsgeschichte*.

[5] "Hellenistic Age" (http://www.britannica.com/EBchecked/topic/260307/Hellenistic-Age). *Encyclopædia Britannica Online*. Encyclopædia Britannica, Inc.. . Retrieved 8 September 2012.

[6] Alexander The Great and the Hellenistic Age, p. xiii. Green P. ISBN 978-0-7538-2413-9

[7] Hellenistic Civilization (http://mars.wnec.edu/~grempel/courses/wc1/lectures/10hellenism.html), Western New England College

[8] Green, Peter. The Hellenistic Age (A Short History). New York: Modern Library Chronicles, 2007.

[9] Jones, Kenneth Raymond (2006). *Provincial reactions to Roman imperialism: the aftermath of the Jewish revolt, A.D. 66-70, Parts 66-70*. University of California, Berkeley. p. 174. ISBN 0-542-82473-6, 9780542824739. "... and the Greeks, or at least the Greco-Macedonian Seleucid Empire, replace the Persians as the Easterners."

[10] Society for the Promotion of Hellenic Studies (London, England) (1993). *The Journal of Hellenic studies, Volumes 113-114*. Society for the Promotion of Hellenic Studies. p. 211. "The Seleucid kingdom has traditionally been regarded as basically a Greco-Macedonian state and its rulers thought of as successors to Alexander."

[11] Baskin, Judith R. ; Seeskin, Kenneth (2010). *The Cambridge Guide to Jewish History, Religion, and Culture*. Cambridge University Press. p. 37. ISBN 0-521-68974-0, 9780521689748. "The wars between the two most prominent Greek dynasties, the Ptolemies of Egypt and the Seleucids of Syria, unalterably change the history of the land of Israel…As a result the land of Israel became part of the empire of the Syrian Greek Seleucids."

[12] Glubb, Sir John Bagot (1967). *Syria, Lebanon, Jordan*. Thames & Hudson. p. 34. OCLC 585939. "In addition to the court and the army, Syrian cities were full of Greek businessmen, many of them pure Greeks from Greece. The senior posts in the civil service were also held by Greeks. Although the Ptolemies and the Seleucids were perpetual rivals, both dynasties were Greek and ruled by means of Greek officials and Greek soldiers. Both governments made great efforts to attract immigrants from Greece, thereby adding yet another racial element to the population."

[13] Steven C. Hause, William S. Maltby (2004). *Western civilization: a history of European society*. Thomson Wadsworth. p. 76. ISBN 0-534-62164-3, 9780534621643. "The Greco-Macedonian Elite. The Seleucids respected the cultural and religious sensibilities of their subjects but preferred to rely on Greek or Macedonian soldiers and administrators for the day-to-day business of governing. The Greek population of the cities, reinforced until the second century BCE by emigration from Greece, formed a dominant, although not especially

cohesive, elite."
[14] Victor, Royce M. (2010). *Colonial education and class formation in early Judaism: a postcolonial reading*. Continuum International Publishing Group. p. 55. ISBN 0-567-24719-8, 9780567247193. "Like other Hellenistic kings, the Seleucids ruled with the help of their "friends" and a Greco-Macedonian elite class separate from the native populations whom they governed."
[15] **Britannica**, *Seleucid kingdom*, 2008, O.Ed.
[16] Life and Land Use on the Bahrain Islands: The Geoarcheology of an Ancient Society By Curtis E. Larsen p. 13
[17] Security and Territoriality in the Persian Gulf: A Maritime Political Geography By Pirouz Mojtahed-Zadeh, page 119
[18] Arnold Hermann Ludwig Heeren, Historical Researches Into the Politics, Intercourse, and Trade of the Principal Nations of Antiquity, Henry Bohn, 1854 p38
[19] Arnold Heeren, ibid, p441
[20] Classical Greece: Ancient histories and modern archaeologies, Ian Morris, Routledge, p184
[21] Phillip Ward, Bahrain: A Travel Guide, Oleander Press p68
[22] W. B. Fisher et al. The Cambridge History of Iran, Cambridge University Press 1968 p40
[23] Jean Francois Salles in Traces of Paradise: The Archaeology of Bahrain, 2500BC-300AD in Michael Rice, Harriet Crawford Ed, IB Tauris, 2002 p132
[24] Jean Francois Salles p132
[25] Curtis E. Larsen. *Life and Land Use on the Bahrain Islands: The Geoarchaeology of an Ancient Society* University Of Chicago Press, 1984 p13
[26] Arnold Heeren, p441
[27] Michael Rice The Archaeology of the Arabian Gulf Routledge 1994 p20
[28] Bahrain By Federal Research Division, page 7
[29] Davies, Cuthbert Collin (1959). *An Historical Atlas of the Indian Peninsula*. Oxford University Press.
[30] Narain, A.K. (1976). *The Coin Types of the Indo-Greek Kings*. Ares. ISBN 0-89005-109-7.
[31] Hans Erich Stier, Georg Westermann Verlag, Ernst Kirsten, and Ekkehard Aner. *Grosser Atlas zur Weltgeschichte: Vorzeit. Altertum. Mittelalter. Neuzeit.* Westermann, 1978, ISBN 3-14-100919-8.
[32] Ghose, Sanujit (2011). "Cultural links between India and the Greco-Roman world" (http://www.ancient.eu.com/article/208/). Ancient History Encyclopedia.
[33] Yavana#cite note-10
[34] Yavana#cite note-11
[35] Yavana#cite note-13
[36] Alexander The Great and the Hellenistic Age, p 98-99. Green, P. ISBN 978-0-7538-2413-9
[37] Alexander The Great and the Hellenistic Age, p 102-103. Green, P. ISBN 978-0-7538-2413-9
[38] Rubicon: Triumph and Tragedy in the Roman Republic. Holland, T. ISBN 978-0-349-11563-4
[39] Cosmos: A Personal Voyage, Sagan, C 1980, "Episode 1: The Shores of the Cosmic Ocean" (http://www.youtube.com/watch?v=Pa1ImgOcOPM&feature=related)
[40] Green, pps. xx, 68-69.
[41] Roy M. MacLeod, *The Library Of Alexandria: Centre Of Learning In The Ancient World*. I.B. Tauris, 2004, ISBN 1-85043-594-4.
[42] Green, p. 21.
[43] Green, p. 23.
[44] Green, p. 22.
[45] Research Machines plc. (2004). *The Hutchinson dictionary of scientific biography*. Abingdon, Oxon: Helicon Publishing. pp. 546. "**Hero of Alexandria (lived c. *AD* 60)** Greek mathematician and engineer, the greatest experimentalist of antiquity"
[46] Lloyd (1973), pp. 1-7.
[47] Lloyd (1973), p. 177.
[48] Otto Neugebauer, *A History of Ancient Mathematical Astronomy*, (New York: Springer, 1975), pp. 284-5; Lloyd (1973), pp. 69-71.
[49] Schaefer Bradley E. (2005). "The Epoch of the Constellations on the Farnese Atlas and Their Origin in Hipparchus's Lost Catalogue". *Journal for the History of Astronomy* **36**: 167–96. Bibcode 2005JHA....36..167S.; But see also Duke Dennis W. (2006). "Analysis of the Farnese Globe". *Journal for the History of Astronomy* **37**: 87–100. Bibcode 2006JHA....37...87D.
[50] Freeth, T.; *et al.* (2006). "Decoding the ancient Greek astronomical calculator known as the Antikythera Mechanism". *Nature* **444** (7119): pp. 587–91. Bibcode 2006Natur.444..587F. doi:10.1038/nature05357. PMID 17136087.; Marchant, Jo (2006). "In Search of Lost Time". *Nature* **444** (7119): pp. 534–8. Bibcode 2006Natur.444..534M. doi:10.1038/444534a. PMID 17136067.;
[51] Charette, François (2006). "High tech from Ancient Greece". *Nature* **444** (7119): pp. 551–2. Bibcode 2006Natur.444..551C. doi:10.1038/444551a. PMID 17136077.; Noble Wilford, John (2006-11-30). "Early Astronomical 'Computer' Found to Be Technically Complex" (http://www.nytimes.com/2006/11/30/science/30compute.html?ex=1322542800&en=088bd939ca75fbbb&ei=5088&partner=rssnyt&emc=rss). *The New York Times*. . Retrieved 2006-11-30.
[52] F. M. Cornford, *The Unwritten Philosophy and Other Essays*, p. 83, quoted in Lloyd (1973), p. 154.
[53] Russo, Lucio (2004). *The Forgotten Revolution: How Science Was Born in 300 BC and Why It Had To Be Reborn*. Berlin: Springer. ISBN 3-540-20396-6. But see the critical reviews by Mott Greene, *Nature*, vol 430, no. 7000 (5 Aug 2004):614 (http://www.nature.com/nature/journal/v430/n7000/full/430614a.html) and Michael Rowan-Robinson, *Physics World*, vol. 17, no. 4 (April 2004) (http://

physicsweb.org/articles/review/17/4/1/1).
[54] Pliny the Elder, *Natural History* (XXXIV, 52)
[55] Green, p. xv.
[56] Herodotus (Holland, T. *Persian Fire*, p. 193.)
[57] Green.

Sources

- Green, Peter. *Alexander The Great and the Hellenistic Age*. Orion Publishing Group, Limited, 2008, ISBN 0-7538-2413-2.

Further reading

- Austin, Michel M., *The Hellenistic world from Alexander to the Roman conquest: a selection of ancient sources in translation*, Cambridge University Press, 1981. ISBN 0-521-22829-8
- Cary, Max, *A History of the Greek World from 323 to 146 B. C.*, London : Methuen & Co. Ltd., 1932
- "Under the Influence: Hellenism in Ancient Jewish Life" (http://www.bib-arch.org/bar/article.asp?PubID=BSBA&Volume=36&Issue=1&ArticleID=29) Biblical Archaeology Society
- Chamoux, F., *Hellenistic Civilization*, 2002
- Tarn, W and Griffith, G. T., *Hellenistic Civilization*, 1952
- Walbank, F. W., *The Hellenistic World*, 1981

External links

- Waterloo Institute for Hellenistic Studies (http://wihs.uwaterloo.ca/)

Mathematician

A **mathematician** is a person with an extensive knowledge of mathematics who uses this knowledge in their work, typically to solve mathematical problems. Mathematics is concerned with numbers, data, collection, quantity, structure, space, and change.

Archimedes was among the greatest mathematicians of antiquity.

Mathematicians involved with solving problems outside of pure mathematics are called applied mathematicians. Applied mathematicians are mathematical scientists who, with their specialized knowledge and professional methodology, approach many of the imposing problems presented in related scientific fields. With professional focus on a wide variety of problems, theoretical systems, and localized constructs, applied mathematicians work regularly in the study and formulation of mathematical models.

The discipline of applied mathematics concerns itself with mathematical methods that are typically used in science, engineering, business, and industry; thus, "applied mathematics" is a mathematical science with specialized knowledge. The term "applied mathematics" also describes the professional specialty in which mathematicians work on problems, often concrete but sometimes abstract. As professionals focused on problem solving, *applied mathematicians* look into the *formulation, study, and use of mathematical models* in science, engineering, business, and other areas of mathematical practice.

Leonhard Euler is widely considered one of the greatest mathematicians.

Education

Mathematicians usually cover a breadth of topics within mathematics in their undergraduate education, and then proceed to specialize in topics of their own choice at the graduate level. In some universities, a qualifying exam serves to test both the breadth and depth of a student's understanding of mathematics; the students who pass are permitted to work on a doctoral dissertation.

Motivation

Mathematicians do research in fields such as logic, set theory, category theory, abstract algebra, number theory, analysis, geometry, topology, dynamical systems, combinatorics, game theory, information theory, numerical analysis, optimization, computation, probability and statistics. These fields comprise both pure mathematics and applied mathematics and establish links between the two. Some fields, such as the theory of dynamical systems, or game theory, are classified as applied mathematics due to the relationships they possess with physics, economics and the other sciences. Whether probability theory and statistics are of theoretical nature, applied nature, or both, is quite controversial among mathematicians. Other branches of mathematics, however, such as logic, number theory, category theory or set theory are accepted as part of pure mathematics, though they find application in other sciences (predominantly computer science and physics). Likewise, analysis, geometry and topology, although considered pure mathematics, find applications in theoretical physics—string theory, for instance.

Carl Friedrich Gauss was the foremost mathematician of the early 19th century.

Though it is true that mathematics finds diverse applications in many areas of research, a mathematician does not determine the value of an idea by the diversity of its applications. Mathematics is interesting in its own right, and a substantial minority of mathematicians investigate the diversity of structures studied in *mathematics itself.* However, among academic mathematics, the majority of mathematical papers published in the United States are written by academics outside of mathematics departments.

Furthermore, a mathematician is not someone who merely manipulates formulas, numbers or equations—the diversity of mathematics allows for research concerning how concepts in one area of mathematics can be used in other areas too. For instance, if one graphs a set of solutions of an equation in some higher dimensional space, he may ask about the geometric properties of the graph. Thus one can understand equations by a pure understanding of abstract topology or geometry—this idea is of importance in algebraic geometry. Similarly, a mathematician does not restrict his study of numbers to the integers; rather he considers more abstract structures such as rings, and in particular number rings in the context of algebraic number theory. This exemplifies the abstract nature of mathematics and how it is not restricted to questions one may ask in daily life.

David Hilbert was one of the most influential mathematicians of the late 19th and early 20th centuries.

In a different direction, mathematicians ask questions about space and transformations that are not restricted to geometric figures such as squares and circles. For instance, an active area of research in differential topology concerns itself with the ways one can "smooth" higher dimensional figures. In fact, whether one can smooth certain higher dimensional spheres was, until recently, an open problem — known as the smooth Poincaré conjecture. Another aspect of mathematics, set-theoretic topology and point-set topology, concerns objects of a different nature from objects in our universe, or in a higher dimensional analogue of our universe. These objects behave in a rather strange manner under deformations, and the properties they possess are completely different from those of objects in our universe. For instance, the "distance" between two points on such an object, may depend on the order in which you consider the pair of points. This is quite different from ordinary life, in which it is accepted that the straight line distance from person A to person B is the same as that between person B and person A.

Emmy Noether is possibly the most influential female mathematician to date.

Another aspect of mathematics, often referred to as "foundational mathematics", consists of the fields of logic and set theory. These explore various ideas regarding the ways one can prove certain claims. This theory is far more complex than it seems, in that the truth of a claim depends on the context in which the claim is made, unlike basic ideas in daily life where truth is absolute. In fact, although some claims may be true, it is impossible to prove or disprove them in rather natural contexts.

Category theory, another field within "foundational mathematics", is rooted on the abstract axiomatization of the definition of a "class of mathematical structures", referred to as a "category". A category intuitively consists of a collection of objects, and defined relationships between them. While these objects may be anything (such as "tables" or "chairs"), mathematicians are usually interested in particular, more abstract, classes of such objects. In any case, it is the *relationships between these objects*, and *not the actual objects* that are predominantly studied.

Differences with scientists

Isaac Newton was a pioneering figure in the development of mathematical physics.

Mathematics differs from natural science in that scientists subject truth claims to tests by experiments, while mathematical propositions are conclusions of mathematical proofs.

If a definite statement is believed plausible by some mathematicians but has been neither proved nor disproved, it is called a *conjecture*, as opposed to an ultimate goal: a *theorem* that has been proved.

Scientific theories change when new information about the world is discovered. Mathematics changes in a similar way: New ideas do not falsify old ones, but new concepts refine old concepts and old theories, attaining a fuller understanding of the truth. One method of refinement is generalization, for example widening the scope of a concept. For instance, calculus (in one variable) generalizes to multivariable calculus, which generalizes to analysis on manifolds. The development of algebraic geometry from its classical to modern forms is a particularly striking example of the way an area of mathematics can change radically in its viewpoint, without making what was correctly proved before in any way incorrect; of course mathematical progress clarifies gaps in previous proofs, often by exposing hidden assumptions, which progress has revealed worth conceptualizing.

A theorem is true, and was true before we knew it and will be true after humans are extinct. Of course, our understanding of what the theorem *really means* gains in profundity as the mathematics around the theorem grows. A mathematician feels that a theorem is better understood when it can be extended to apply in a broader setting than previously known. For instance, Fermat's little theorem for the nonzero integers modulo a prime generalizes to Euler's theorem for the invertible numbers modulo any nonzero integer, which generalizes to Lagrange's theorem for finite groups.

Women in mathematics

While the majority of mathematicians are male, there have been some demographic changes since World War II. Some prominent female mathematicians are Hypatia of Alexandria (ca. 400 AD), Ada Lovelace (1815–1852), Maria Gaetana Agnesi (1718–1799), Emmy Noether (1882–1935), Sophie Germain (1776–1831), Sofia Kovalevskaya (1850–1891), Alicia Boole Stott (1860–1940), Rózsa Péter (1905–1977), Julia Robinson (1919–1985), Olga Taussky-Todd (1906–1995), Émilie du Châtelet (1706–1749), Mary Cartwright (1900–1998), Olga Ladyzhenskaya (1922–2004), and Olga Oleinik (1925–2001).

The Association for Women in Mathematics is a professional society whose purpose is "to encourage women and girls to study and to have active careers in the mathematical sciences, and to promote equal opportunity and the equal treatment of women and girls in the mathematical sciences." The American Mathematical Society and other mathematical societies offer several prizes aimed at increasing the representation of women and minorities in the future of mathematics.

Prizes in mathematics

There is no Nobel Prize in mathematics, though sometimes mathematicians have won the Nobel Prize in a different field, such as economics. Prominent prizes in mathematics include the Abel Prize, the Chern Medal, the Fields Medal, the Gauss Prize, the Nemmers Prize, the Balzan Prize, the Crafoord Prize, the Shaw Prize, the Steele Prize, the Wolf Prize, the Schock Prize, and the Nevanlinna Prize.

Quotations about mathematicians

The following are quotations about mathematicians, or by mathematicians.

> *A mathematician is a device for turning coffee into theorems.*
>
> —Attributed to both Alfréd Rényi[1] and Paul Erdős

> *Die Mathematiker sind eine Art Franzosen; redet man mit ihnen, so übersetzen sie es in ihre Sprache, und dann ist es alsobald ganz etwas anderes.* (Mathematicians are [like] a sort of Frenchmen; if you talk to them, they translate it into their own language, and then it is immediately something quite different.)
>
> —Johann Wolfgang von Goethe

> *Each generation has its few great mathematicians...and [the others'] research harms no one.*
>
> —Alfred W. Adler (1930-), "Mathematics and Creativity"[2]

> *In short, I never yet encountered the mere mathematician who could be trusted out of equal roots, or one who did not clandestinely hold it as a point of his faith that x squared + px was absolutely and unconditionally equal to q. Say to one of these gentlemen, by way of experiment, if you please, that you believe occasions may occur where x squared + px is not altogether equal to q, and, having made him understand what you mean, get out of his reach as speedily as convenient, for, beyond doubt, he will endeavor to knock you down.*
>
> —Edgar Allan Poe, *The purloined letter*

> *A mathematician, like a painter or poet, is a maker of patterns. If his patterns are more permanent than theirs, it is because they are made with ideas.*
>
> —G. H. Hardy, *A Mathematician's Apology*

> *Some of you may have met mathematicians and wondered how they got that way.*
>
> —Tom Lehrer

> *It is impossible to be a mathematician without being a poet in soul.*
>
> —Sofia Kovalevskaya

> *There are two ways to do great mathematics. The first is to be smarter than everybody else. The second way is to be stupider than everybody else—but persistent.*
>
> —Raoul Bott

See also

Some notable mathematicians include Archimedes of Syracuse, Leonhard Euler, Carl Gauss, Johann Bernoulli, Jacob Bernoulli, Aryabhata, Brahmagupta, Bhaskara II, Nilakantha Somayaji, Omar Khayyám, Muhammad ibn Mūsā al-Khwārizmī, Bernhard Riemann, Gottfried Leibniz, Andrey Kolmogorov, Euclid of Alexandria, Jules Henri Poincaré, Srinivasa Ramanujan, Alexander Grothendieck, David Hilbert, Alan Turing, von Neumann, Kurt Gödel, Joseph Louis Lagrange, Georg Cantor, William Rowan Hamilton, Carl Jacobi, Évariste Galois, Nikolay Lobachevsky, Rene Descartes, Joseph Fourier, Pierre-Simon Laplace, Alonzo Church, Nikolay Bogolyubov and Pierre de Fermat.

- List of amateur mathematicians

- List of female mathematicians
- List of mathematicians
- Mathematical joke
- *Men of mathematics* (book)
- *A Mathematician's Apology*

Notes

[1] "Biography of Alfréd Rényi" (http://www-history.mcs.st-andrews.ac.uk/Biographies/Renyi.html). History.mcs.st-andrews.ac.uk. . Retrieved 2012-08-17.

[2] Alfred Adler, "Mathematics and Creativity," *The New Yorker*, 1972, reprinted in Timothy Ferris, ed., *The World Treasury of Physics, Astronomy, and Mathematics*, Back Bay Books, reprint, June 30, 1993, p, 435.

References

- *A Mathematician's Apology*, by G. H. Hardy. Memoir, with foreword by C. P. Snow.
  - Reprint edition, Cambridge University Press, 1992; ISBN 0-521-42706-1
  - First edition, 1940
- Paul Halmos. *I Want to Be a Mathematician*. Springer-Verlag 1985.
- Dunham, William. *The Mathematical Universe*. John Wiley 1994.

External links

- Occupational Outlook: Mathematicians (http://stats.bls.gov/oco/ocos043.htm). Information on the occupation of mathematician from the US Department of Labor.
- Sloan Career Cornerstone Center: Careers in Mathematics (http://www.careercornerstone.org/math/math.htm). Although US-centric, a useful resource for anyone interested in a career as a mathematician. Learn what mathematicians do on a daily basis, where they work, how much they earn, and more.
- The MacTutor History of Mathematics archive (http://www-history.mcs.st-and.ac.uk/history/index0.html). A comprehensive list of detailed biographies.
- The Mathematics Genealogy Project (http://genealogy.math.ndsu.nodak.edu/). Allows to follow the succession of thesis advisors for most mathematicians, living or dead.
- Weisstein, Eric W., " Unsolved Problems (http://mathworld.wolfram.com/UnsolvedProblems.html)" from MathWorld.
- Middle School Mathematician Project (http://valure.wiki.ccsd.edu/) Short biographies of select mathematicians assembled by middle school students.
- Career Information for Students of Math and Aspiring Mathematicians (http://www.mathmajor.org/careers) from MathMajor (http://mathmajor.org)

Theon_of_Smyrna

Theon of Smyrna (Θέωνος ὁ Σμυρναῖος) (fl. 100 CE) was a Greek philosopher and mathematician, whose works were strongly influenced by the Pythagorean school of thought. His surviving *On Mathematics Useful for the Understanding of Plato* is an introductory survey of Greek mathematics.

Life

Little is known about the life of Theon of Smyrna. A bust created at his death, and dedicated by his son, was discovered at Smyrna, and art historians date it to around 135 CE. Ptolemy refers several times in his *Almagest* to a Theon who made observations at Alexandria, but it is uncertain whether he is referring to Theon of Smyrna.[1] The lunar impact crater Theon Senior is named for him.

Works

Theon wrote several commentaries on the works of mathematicians and philosophers of the time, including works on the philosophy of Plato. Most of these works are lost. The one major survivor is his *On Mathematics Useful for the Understanding of Plato*. A second work concerning the order in which to study Plato's works has recently been discovered in an Arabic translation.[2]

On Mathematics Useful for the Understanding of Plato

His *On Mathematics Useful for the Understanding of Plato* is not a commentary on Plato's writings but rather a general handbook for a student of mathematics. It is not so much a groundbreaking work as a reference work of ideas already known at the time. Its status as a compilation of already-established knowledge and its thorough citation of earlier sources is part of what makes it valuable.

The first part of this work is divided into two parts, the first covering the subjects of numbers and the second dealing with music and harmony. The first section, on mathematics, is most focused on what today is most commonly known as number theory: odd numbers, even numbers, prime numbers, perfect numbers, abundant numbers, and other such properties. It contains an account of 'side and diameter numbers', the sequence of best rational approximations to the square root of 2, [3] the denominators of which are Pell numbers.

The second section, on music, is split into three parts: music of numbers (*hē en arithmois mousikē*), instrumental music (*hē en organois mousikē*), and "music of the spheres" (*hē en kosmō harmonia kai hē en toutō harmonia*). The "music of numbers" is a treatment of temperament and harmony using ratios, proportions, and means; the sections on instrumental music concerns itself not with melody but rather with intervals and consonances in the manner of Pythagoras' work. Theon considers intervals by their degree of consonance: that is, by how simple their ratios are. (For example, the octave is first, with the simple 2:1 ratio of the octave to the fundamental.) He also considers them by their distance from one another.

The third section, on the music of the cosmos, he considered most important, and ordered it so as to come after the necessary background given in the earlier parts. Theon quotes a poem by Alexander of Ephesus assigning specific pitches in the chromatic scale to each planet, an idea that would retain its popularity for a millennium thereafter.

The second book is on astronomy. Here Theon affirms the spherical shape and large size of the Earth; he also describes the occultations, transits, conjunctions, and eclipses. However, the quality of the work led Otto Neugebauer to criticize him for not fully understanding the material he attempted to present

Notes

[1] James Evans, (1998), *The History and Practice of Ancient Astronomy*, page 49

[2] "Theon of Smyrna" entry in John Hazel, 2002, *Who's who in the Greek world*, page 37. Routledge

[3] http://books.google.co.uk/books?id=dmY3Vjix3kC&pg=PA112&lpg=PA112&dq=side+and+diameter+numbers+theon&source=bl&ots=v1_UAqyie5&sig=z0FkNP5tcGHCoc-DWKf5v40KotE&hl=en&sa=X&ei=JrqDUMvTCMi_0QWu6ICoBA&ved=0CB0Q6AEwAA#v=onepage&q=side%20and%20diameter%20numbers%20theon&f=false

References

- Theon, of Smyrna: *Mathematics useful for understanding Plato; translated from the 1892 Greek/French edition of J. Dupuis by Robert and Deborah Lawlor and edited and annotated by Christos Toulis and others; with an appendix of notes by Dupuis, a copious glossary, index of works, etc.* Series: *Secret doctrine reference series* Published: San Diego : Wizards Bookshelf, 1979. ISBN 0-913510-24-6. 174pp.
- E.Hiller, Theonis Smyrnaei: expositio rerum mathematicarum ad legendum Platonem utilium (http://books.google.com/books?id=6r0UAAAAQAAJ), Leipzig:Tuebner, 1878, repr. 1966.
- J. Dupuis, Exposition des connaissances mathematiques utiles pour la lecture de platon (http://books.google.com/books?id=8RdKAAAAMAAJ), 1892. French translation.
- Lukas Richter:"Theon of Smyrna". Grove Music Online, ed. L. Macy. Accessed 29 Jun 05. (subscription access) (http://www.grovemusic.com)
- O'Connor, John J.; Robertson, Edmund F., "Theon of Smyrna" (http://www-history.mcs.st-andrews.ac.uk/Biographies/Theon_of_Smyrna.html), *MacTutor History of Mathematics archive*, University of St Andrews.

Hadrian

Hadrian	
14th Emperor of the Roman Empire	
 Marble bust of Hadrian at the Palazzo dei Conservatori, Capitoline Museums.	
Reign	10 August 117 – 10 July 138
Full name	Publius Aelius Hadrianus Buccellanus (from birth to adoption and accession); Caesar Publius Aelius Traianus Hadrianus Buccellanus Augustus (as emperor)
Born	24 January 76
Birthplace	Italica, Spain; or Rome (uncertain)
Died	10 July 138 (aged 62)
Place of death	Baiae
Buried	1) Puteoli 2) Gardens of Domitia 3) Hadrian's Mausoleum (Rome)
Predecessor	Trajan
Successor	Antoninus Pius
Wife	Vibia Sabina
Offspring	Lucius Aelius, Antoninus Pius (both adoptive)
Dynasty	Nervan-Antonine
Father	Publius Aelius Hadrianus Afer
Mother	Domitia Paulina

Roman imperial dynasties

Nervo-Trajanic Dynasty

Nerva
Children
Natural - (none)
Adoptive - Trajan
Trajan
Children
Natural - (none)
Adoptive - Hadrian
Hadrian
Children
Natural - (none)
Adoptive - Lucius Aelius
Adoptive - Antoninus Pius

Hadrian (Latin: *Publius Aelius Traianus Hadrianus Augustus*[1] [2] [3] 24 January, 76 AD – 10 July, 138 AD), was Roman Emperor from 117 to 138. He is best known for building Hadrian's Wall, which marked the northern limit of Roman Britain. In Rome, he re-built the Pantheon and constructed the Temple of Venus and Roma. In addition to being emperor, Hadrian was a humanist and was philhellene in most of his tastes. He was the third of the so-called Five Good Emperors.

Hadrian was born Publius Aelius Hadrianus to an ethnically Italian family, either in Italica near Seville (in modern-day Spain) or in Rome. His predecessor Trajan, also Hispanic himself, was a maternal cousin of Hadrian's father.[4] Trajan never officially designated an heir, but according to his wife Pompeia Plotina, Trajan named Hadrian emperor immediately before his death. Trajan's wife and his friend Licinius Sura were well-disposed towards Hadrian, and he may well have owed his succession to them.[5]

During his reign, Hadrian traveled to nearly every province of the Empire. An ardent admirer of Greece, he sought to make Athens the cultural capital of the Empire and ordered the construction of many opulent temples in the city. He used his relationship with his Greek favorite Antinous to underline his philhellenism and led to the creation of one of the most popular cults of ancient times. He spent extensive amounts of his time with the military; he usually wore military attire and even dined and slept amongst the soldiers. He ordered military training and drilling to be more rigorous and even made use of false reports of attack to keep the army alert.

Upon his accession to the throne, Hadrian withdrew from Trajan's conquests in Mesopotamia and Armenia, and even considered abandoning Dacia. Late in his reign he suppressed the Bar Kokhba revolt in Judaea, renaming the province Syria Palaestina. In 136 an ailing Hadrian adopted Lucius Aelius as his heir, but the latter died suddenly two years later. In 138, Hadrian resolved to adopt Antoninus Pius if he would in turn adopt Marcus Aurelius and Aelius' son Lucius Verus as his own eventual successors. Antoninus agreed, and soon afterward Hadrian died at Baiae.[6]

Early life

Hadrian was born Publius Aelius Hadrianus in Italica,[7] or in Rome,[8] from a well-established family which had originated in Picenum in Italy and had subsequently settled in Italica, Hispania Baetica (the republican Hispania Ulterior), near the present-day location of Seville, Spain.

Although it was an accepted part of Hadrian's personal history that he was born in Spain, his biography in Augustan History states that he was born in Rome on 24 January 76, of a family originally Italian, but which had lived in Spain for many generations. However, this may be a ruse to make Hadrian look like a person from Rome instead of a person hailing from the provinces.[9] His father was Publius Aelius Hadrianus Afer, who as a senator of praetorian rank would spend much of his time in Rome.[10]

Hadrian's forefathers came from Hadria, modern Atri, an ancient town of Picenum in Italy, but the family had settled in Italica in Hispania Baetica soon after its founding by Scipio Africanus. Afer was a paternal cousin of the future Emperor Trajan. His mother was Domitia Paulina who came from Gades (Cádiz). Paulina was a daughter of a distinguished Spanish-Roman Senatorial family.[11]

Hadrian's elder sister and only sibling was Aelia Domitia Paulina, married with the triple consul Lucius Julius Ursus Servianus, his niece was Julia Serviana Paulina and his great-nephew was Gnaeus Pedanius Fuscus Salinator, from Barcino (Barcelona). His parents died in 86 when Hadrian was ten, and the boy then became a ward of both Trajan and Publius Acilius Attianus (who was later Trajan's Praetorian Prefect).[11] Hadrian was schooled in various subjects particular to young aristocrats of the day, and was so fond of learning Greek literature that he was nicknamed *Graeculus* ("Greekling").

Hadrian visited Italica when (or never left it until) he was 14 years old, when he was recalled by Trajan, who thereafter looked after his development. He never returned to Italica although it was later made a colonia in his honour.[12]

Public service

His first military service was as a tribune of the Legio II *Adiutrix*. Later, he was to be transferred to the Legio I *Minervia* in Germany. When Nerva died in 98, Hadrian rushed to inform Trajan personally. He later became legate of a legion in Upper Pannonia and eventually governor of said province. He was also archon in Athens for a brief time, and was elected an Athenian citizen.[13]

His career before becoming emperor follows:

- *decemvir stlitibus iudicandis*
- *sevir turmae equitum Romanorum*
- *praefectus Urbi feriarum Latinarum*
- *tribunus militum legionis II Adiutricis Piae Fidelis* (95, in Pannonia Inferior)
- *tribunus militum legionis V Macedonicae* (96, in Moesia Inferior)
- *tribunus militum legionis XXII Primigeniae Piae Fidelis* (97, in Germania Superior)
- *quaestor* (101)
- *ab actis senatus*
- *tribunus plebis* (105)
- *praetor* (106)
- *legatus legionis I Minerviae Piae Fidelis* (106, in Germania Inferior)
- *legatus Augusti pro praetore Pannoniae Inferioris* (107)
- *consul suffectus* (108)
- *septemvir epulonum* (before 112)
- *sodalis Augustalis* (before 112)
- *archon Athenis* (112/13)

- *legatus Syriae* (117).[14]

Hadrian was involved in the wars against the Dacians (as legate of the V *Macedonica*) and reputedly won awards from Trajan for his successes. Hadrian's military skill is not well-attested due to a lack of military action during his reign; however, his keen interest and knowledge of the army and his demonstrated skill of leadership show possible strategic talent.

Hadrian joined Trajan's expedition against Parthia as a legate on Trajan's staff.[15] Neither during the first victorious phase, nor during the second phase of the war when rebellion swept Mesopotamia did Hadrian do anything of note. However, when the governor of Syria had to be sent to sort out renewed troubles in Dacia, Hadrian was appointed as a replacement, giving him an independent command.[16]

Trajan, seriously ill by that time, decided to return to Rome while Hadrian remained in Syria to guard the Roman rear. Trajan only got as far as Selinus before he became too ill to go further. While Hadrian may have been the obvious choice as successor, he had never been adopted as Trajan's heir. As Trajan lay dying, nursed by his wife, Plotina (a supporter of Hadrian), he at last adopted Hadrian as heir. Since the document was signed by Plotina, it has been suggested that Trajan may have already been dead.[17]

Emperor (117)

Securing power

Hadrian quickly secured the support of the legions — one potential opponent, Lusius Quietus, was promptly dismissed.[18] The Senate's endorsement followed when possibly falsified papers of adoption from Trajan were presented (although he had been the ward of Trajan). The rumour of a falsified document of adoption carried little weight — Hadrian's legitimacy arose from the endorsement of the Senate and the Syrian armies.

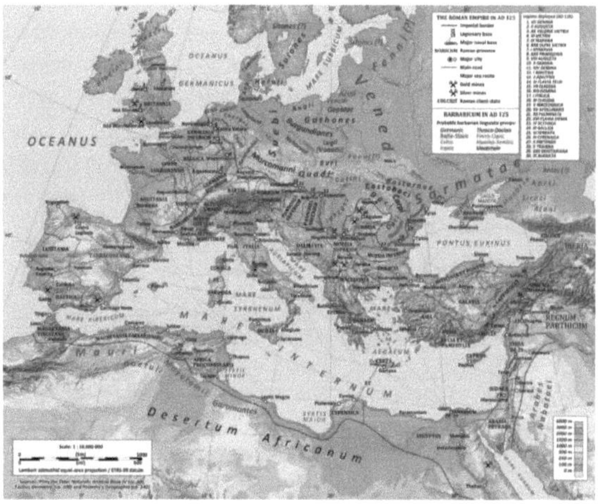

The Roman empire in 125, under the rule of Hadrian.

Hadrian did not at first go to Rome — he was busy sorting out the East and suppressing the Jewish revolt that had broken out under Trajan, then moving on to sort out the Danube frontier. Instead, Attianus, Hadrian's former guardian, was put in charge in Rome. There he "discovered" a conspiracy involving four leading Senators including Lusius Quietus and demanded of the Senate their deaths.[19]

There was no question of a trial—they were hunted down and killed out of hand. Because Hadrian was not in Rome at the time, he was able to claim that Attianus had acted on his own initiative. According to Elizabeth Speller, the real reason for their deaths was that they were Trajan's men.[19]

Hadrian and the military

Despite his own great reputation as a military administrator, Hadrian's reign was marked by a general lack of major military conflicts, apart from the Second Roman-Jewish War. He surrendered Trajan's conquests in Mesopotamia, considering them to be indefensible. There was almost a war with Parthia around 121, but the threat was averted when Hadrian succeeded in negotiating a peace.

The peace policy was strengthened by the erection of permanent fortifications along the empire's borders (*limites*, sl. *limes*). The most famous of these is the massive Hadrian's Wall in Great Britain, and the Danube and Rhine borders were strengthened with a series of mostly wooden fortifications, forts, outposts and watchtowers, the latter specifically improving communications and local area security. To maintain morale and prevent the troops from becoming restive, Hadrian established intensive drill routines, and personally inspected the armies. Although his coins showed military images almost as often as peaceful ones, Hadrian's policy was peace through strength, even threat.[20]

Antinous

Hadrian had a close relationship with a Bithynian Greek youth, Antinous, which was most likely sexual.[21] Edward Gibbon remarks, "[O]f the first fifteen emperors, Claudius was the only one whose taste in love was entirely correct."[22] According to *Historia Augusta*:

> During a journey on the Nile he lost Antinous, his favourite, and for this youth he wept like a woman. Concerning this incident there are varying rumours; for some claim that he had devoted himself to death for Hadrian, and others — what both his beauty and Hadrian's sensuality suggest. But however this may be, the Greeks deified him at Hadrian's request, and declared that oracles were given through his agency, but these, it is commonly asserted, were composed by Hadrian himself.[23]

Antinous drowned in 130. Deeply saddened, Hadrian founded the Egyptian city of Antinopolis in his memory, and had Antinous deified – an unprecedented honour for one not of the ruling family. The cult of Antinous became very popular in the Greek-speaking world.[24] It has been suggested that Hadrian created the cult as a political move to reconcile the Greek-speaking East to Roman rule.[25]

Cultural pursuits and patronage

Hadrian has been described, by Ronald Syme among others, as the most versatile of all the Roman Emperors. He also liked to demonstrate knowledge of all intellectual and artistic fields. Above all, Hadrian patronized the arts: Hadrian's Villa at Tibur (Tivoli) was the greatest Roman example of an Alexandrian garden, recreating a sacred landscape, lost in large part to the despoliation of the ruins by the Cardinal d'Este who had much of the marble removed to build Villa d'Este. In Rome, the Pantheon, originally built by Agrippa but destroyed by fire in 80, was rebuilt under Hadrian in the domed form it retains to this day. It is among the best preserved of Rome's ancient buildings and was highly influential to many of the great architects of the Italian Renaissance and Baroque periods.

From well before his reign, Hadrian displayed a keen interest in architecture, but it seems that his eagerness was not always well received. For example, Apollodorus of Damascus, famed architect of the Forum of Trajan, dismissed his designs. When Trajan, predecessor to Hadrian, consulted Apollodorus about an architectural problem, Hadrian interrupted to give advice, to which Apollodorus replied, "Go away and draw your pumpkins. You know nothing about these problems." "Pumpkins" refers to Hadrian's drawings of domes like the Serapeum in his villa. It is rumoured that once Hadrian succeeded Trajan to become emperor, he had Apollodorus exiled and later put to death. It is very possible that this later story was a later attempt to defame his character.

Hadrian wrote poetry in both Latin and Greek; one of the few surviving examples is a Latin poem he reportedly composed on his deathbed (see below). He also wrote an autobiography – not, apparently, a work of great length or revelation, but designed to scotch various rumours or explain his various actions. Hadrian was a passionate hunter from the time of his youth according to one source.[26] In northwest Asia, he founded and dedicated a city to commemorate a she-bear he killed.[27] It is documented that in Egypt he and his beloved Antinous killed a lion.[27] In Rome, eight reliefs featuring Hadrian in different stages of hunting decorate a building that began as a monument celebrating a kill.[27]

Another of Hadrian's contributions to "popular" culture was the beard, which symbolised his philhellenism. Since the time of Scipio Africanus it had been fashionable among the Romans to be clean-shaven. Also all Roman Emperors before Hadrian, except for Nero (also a great admirer of Greek culture), were clean shaven. Most of the emperors after Hadrian would be portrayed with beards. Their beards, however, were not worn out of an appreciation for Greek culture but because the beard had, thanks to Hadrian, become fashionable. This new fashion lasted until the reign of Constantine the Great[28] and was revived again by Phocas at the start of the 7th century.[29]

As a cultural Hellenophile Hadrian was familiar with the work of the philosophers Epictetus, Heliodorus and Favorinus. At home he attended to social needs. Hadrian mitigated but did not abolish slavery, had the legal code humanized and forbade torture. He built libraries, aqueducts, baths and theaters. Hadrian is considered by many historians to have been wise and just: Schiller called him "the Empire's first servant", and British historian Edward Gibbon admired his "vast and active genius", as well as his "equity and moderation". In 1776, he stated that Hadrian's era was part of the "happiest era of human history".

While visiting Greece in 126, Hadrian attempted to create a kind of provincial parliament to bind all the semi-autonomous former city states across all Greece and Ionia (in Asia Minor). This parliament, known as the Panhellenion, failed despite spirited efforts to foster cooperation among the Hellenes. Hadrian died at his villa in Baiae. He was buried in a mausoleum on the western bank of the Tiber, in Rome, a building later transformed into a papal fortress, Castel Sant'Angelo. The dimensions of his mausoleum, in its original form, were deliberately designed to be slightly larger than the earlier Mausoleum of Augustus.

According to Cassius Dio a gigantic equestrian statue was erected to Hadrian after his death. "It was so large that the bulkiest man could walk through the eye of each horse, yet because of the extreme height of the foundation persons passing along on the ground below believe that the horses themselves as well as Hadrian are very very small."

Hadrian's travels

Purpose

The Emperor travelled broadly, inspecting and correcting the legions in the field. Even prior to becoming emperor, he had travelled abroad with the Roman military, giving him much experience in the matter. More than half his reign was spent outside of Italy. Other emperors often left Rome simply to go to war, returning soon after conflicts concluded. A previous emperor, Nero, once travelled through Greece and was condemned for his self-indulgence.

Hadrian, by contrast, travelled as a fundamental part of his governing, and made this clear to the Roman Senate and the people. He was able to do this because at Rome he possessed a staunch supporter within the upper echelons of Roman society, a military veteran by the name of Marcius Turbo. Also, there are hints within certain sources that he also employed a secret police force, the frumentarii, to exert control and influence in case anything should go wrong while he journeyed abroad.

His visits were marked by handouts which often contained instructions for the construction of new public buildings. His intention was to strengthen the Empire from within through improved infrastructure, as opposed to conquering or annexing perceived enemies. This was often the purpose of his journeys; commissioning new structures, projects and settlements. His almost evangelical belief in Greek culture strengthened his views: like many emperors before him, Hadrian's will was almost always obeyed.[30]

His travelling court was large, including administrators and likely architects and builders. The burden on the areas he passed through were sometimes great. While his arrival usually brought some benefits it is possible that those who had to bear the burden were of different class to those who reaped the benefits. For example, huge amounts of provisions were requisitioned during his visit to Egypt, this suggests that the burden on the mainly subsistence farmers must have been intolerable, causing some measure of starvation and hardship.[30]

At the same time, as in later times all the way through the European Renaissance, kings were welcomed into their cities or lands, and the financial burden was completely on them, and only indirectly on the poorer class. Hadrian's first tour came in 121 and was initially aimed at covering his back to allow himself the freedom to concern himself with his general cultural aims. He travelled north, towards Germania and inspected the Rhine-Danube frontier, allocating funds to improve the defences. However, it was a voyage to the Empire's very frontiers that represented his perhaps most significant visit; upon hearing of a recent revolt, he journeyed to Britannia.

Britannia

Prior to Hadrian's arrival on Great Britain there had been a major rebellion in Britannia from 119 to 121.[31] In 122 he initiated the construction of Hadrian's Wall. The wall was built, "to separate Romans from barbarians," according to the *Historia Augusta.* Augustan Histories [32] It deterred attacks on Roman territory and controlled cross border trade and immigration.[33]

Hadrian's Wall (Vallum Hadriani), a fortification in Northern England (viewed from Vercovicium).

Unlike the Germanic limes, built of wood palisades, the lack of suitable wood in the area required a stone construction.[34] The western third of the wall, from modern-day Carlisle to the River Irthing, was built of turf because of the lack of suitable building stone. This problem also led to the narrowing of the width of the wall, from the original 12 feet to 7.[35]

Under him, a shrine was erected in York to Britain as a Goddess, and coins that introduced a female figure as the personification of Britain, labeled BRITANNIA, were struck.[36] By the end of 122, he had concluded his visit to Britannia, and from there headed south by sea to Mauretania.

Hadrian's Gate, in Antalya, southern Turkey was built to honour Hadrian who visited the city in 130.

Parthia and Anatolia

In 123, he arrived in Mauretania where he personally led a campaign against local rebels.[37] However, this visit was to be short, as reports came through that the Eastern nation of Parthia was again preparing for war; as a result, Hadrian quickly headed eastwards. On his journey east it is known that at some point he visited Cyrene during which he personally made available funds for the training of the young men of well bred families for the Roman military. This might well have been a stop off during his journey East. Cyrene had already benefited from his generosity when he in 119 had provided funds for the rebuilding of public buildings destroyed in the recent Jewish revolt.[38]

When Hadrian arrived on the Euphrates, he characteristically solved the problem through a negotiated settlement with the Parthian king Osroes I. He then proceeded to check the Roman defences before setting off West along the coast of the Black Sea.[39] He probably spent the winter in Nicomedia, the main city of Bithynia. As Nicomedia had been hit by an earthquake only shortly prior to his stay, Hadrian was generous in providing funds for rebuilding. Thanks to his generosity he was acclaimed as the chief restorer of the province as a whole.[40]

It is more than possible that Hadrian visited Claudiopolis and there espied the beautiful Antinous, a young boy who was destined to become the emperor's beloved. Sources say nothing about when Hadrian met Antinous; however, there are depictions of Antinous that shows him as a young man of 20 or so. As this was shortly before Antinous's drowning in 130, Antinous would most likely have been a youth of 13 or 14.[40] It is possible that Antinous may have been sent to Rome to be trained as a page to serve the emperor, and only gradually did he rise to the status of imperial favourite.[41]

After meeting Antinous, Hadrian travelled through Anatolia. The route he took is uncertain. Various incidents are described, such as his founding of a city within Mysia, Hadrianutherae, after a successful boar hunt. (The building of the city was probably more than a mere whim — low-populated wooded areas such as the location of the new city were already ripe for development). Some historians dispute whether Hadrian did in fact commission the city's construction at all. At about this time, plans to build a temple in Asia Minor were written up. The new temple would be dedicated to Trajan and Hadrian, built with dazzling white marble.[42]

Greece

The climax of this tour was the destination that the hellenophile Hadrian must all along have had in mind, Greece. He arrived in the autumn of 124 in time to participate in the Eleusinian Mysteries. By tradition, at one stage in the ceremony the initiates were supposed to carry arms; but, this was waived to avoid any risk to the emperor. At the Athenians' request, he conducted a revision of their constitution — among other things a new phyle (tribe) was added bearing his name.[43]

Temple of Zeus in Athens.

During the winter he toured the Peloponnese. His exact route is uncertain; however, Pausanias reports of tell-tale signs, such as temples built by Hadrian and the statue of the emperor built by the grateful citizens of Epidaurus in thanks to their "restorer". He was especially generous to Mantinea; this supports the theory that Antinous was in fact already Hadrian's lover because of the strong link between Mantinea and Antinous's home in Bithynia.[44]

The Pantheon was rebuilt by Hadrian.

By March 125, Hadrian had reached Athens. presiding over the festival of Dionysia. The building program that Hadrian initiated was substantial. Various rulers had done work on building the Temple of Olympian Zeus over a timespan of more than five centuries — it was Hadrian and the vast resources he could command that ensured that the job would be finished. He also initiated the construction of several public buildings on his own whim and even organized the building of an aqueduct.[45]

Return to Italy

On his return to Italy, Hadrian made a detour to Sicily. Coins celebrate him as the restorer of the island, though there is no record of what he did to earn this accolade.[46]

Back in Rome, he was able to see for himself the completed work of rebuilding the Pantheon. Also completed by then was Hadrian's villa nearby at Tibur, a pleasant retreat by the Sabine Hills for whenever Rome became too much for him. At the beginning of March 127 Hadrian set off for a tour of Italy. Once again, historians are able to reconstruct his route by evidence of his hand-outs rather than the historical records.[47]

For instance, in that year he restored the Picentine earth goddess Cupra in the town of Cupra Maritima. At some unspecified time he improved the drainage of the Fucine lake. Less welcome than such largesse was his decision to divide Italy into 4 regions under imperial legates with consular rank. Being effectively reduced to the status of mere provinces did not go down well and this innovation did not long outlive Hadrian.[47]

Hadrian fell ill around this time, though the nature of his sickness is not known. Whatever the illness was, it did not stop him from setting off in the spring of 128 to visit Africa. His arrival began with the good omen of rain ending a drought. Along with his usual role as benefactor and restorer, he found time to inspect the troops; his speech to the troops survives to this day.[48] Hadrian returned to Italy in the summer of 128 but his stay was brief, as he set off on another tour that would last three years.[49]

Greece, Asia, and Egypt (128-130)

In September 128, Hadrian again attended the Eleusinian mysteries. This time his visit to Greece seems to have concentrated on Athens and Sparta — the two ancient rivals for dominance of Greece. Hadrian had played with the idea of focusing his Greek revival round Amphictyonic League based in Delphi, but he by now had decided on something far grander. His new Panhellenion was going to be a council that would bring Greek cities together wherever they might be found. The meeting place was to be the new temple to Zeus in Athens. Having set in motion the preparations — deciding whose claim to be a Greek city was genuine would in itself take time — Hadrian set off for Ephesus.[50]

In October 130, while Hadrian and his entourage were sailing on the Nile, Antinous drowned for unknown reasons; accident, suicide, murder or religious sacrifice have all been postulated. The emperor was grief-stricken. He ordered Antinous deified, and cities were named after the boy, medals struck with his effigy, and statues erected to him in all parts of the empire. Temples were built for his worship in Bithynia, Mantineia in Arcadia. In Athens, festivals were celebrated in his honour and oracles delivered in his name. The city of Antinopolis or Antinoe was founded on the ruins of Besa, where he died.[51]

Greece, Judaea, and Illyricum

Hadrian's movements subsequent to the founding of Antinopolis on 30 October 130 are obscure. Whether or not he returned to Rome, he spent the winter of 131–32 in Athens and probably remained in Greece or further East because of the Jewish rebellion which broke out in Judaea in 132 (see below). Inscriptions make it clear that he took the field in person against the rebels with his army in 133; he then returned to Rome, probably in that year and almost certainly (judging again from inscriptions) via Illyricum.[52]

Second Roman-Jewish War (132–136)

In 130, Hadrian visited the ruins of Jerusalem, in Judaea, left after the First Roman-Jewish War of 66–73. He rebuilt the city, renaming it Aelia Capitolina after himself and Jupiter Capitolinus, the chief Roman deity. Hadrian placed the city's main Forum at the junction of the main Cardo and Decumanus Maximus, now the location for the (smaller) Muristan. Hadrian built a large temple to the goddess Venus on top of what later Christians would come to venerate as the tomb of Christ.[53]

Later, during the reign of Constantine, this site was destroyed and rebuilt as the Church of the Holy Sepulchre after the Christian Empress Helena ordered the temple of Venus to be demolished.[53] A new temple dedicated to the worship of Jupiter was built on the ruins of the old Jewish Second Temple, which had been destroyed in 70.[54] In addition, Hadrian abolished circumcision, which was considered by Romans and Greeks as a form of bodily mutilation and hence barbaric.[55]

These anti-Jewish policies of Hadrian triggered in Judaea a massive Jewish uprising, led by Simon bar Kokhba and Akiba ben Joseph. Following the outbreak of the revolt, Hadrian called his general Sextus Julius Severus from Britain, and troops were brought from as far as the Danube. Roman losses were very heavy, and it is believed that an entire legion, the XXII Deiotariana was destroyed.[56] Indeed, Roman losses were so heavy that Hadrian's report to the Roman Senate omitted the customary salutation "I and the legions are well".[57]

Hadrian's army eventually put down the rebellion in 135, after three years of fighting. According to Cassius Dio, during the war 580,000 Jews were killed, 50 fortified towns and 985 villages razed. The final battle took place in Beitar, a fortified city 10 km. southwest of Jerusalem. The city only fell after a lengthy siege, and Hadrian only allowed the Jews to bury their dead after a period of six days. According to the Babylonian Talmud,[58] after the war Hadrian continued the persecution of Jews.

He attempted to root out Judaism, which he saw as the cause of continuous rebellions, prohibited the Torah law, the Hebrew calendar and executed Judaic scholars (see Ten Martyrs). The sacred scroll was ceremonially burned on the Temple Mount. In an attempt to erase the memory of Judaea, he renamed the province Syria Palaestina (after the Philistines), and Jews were forbidden from entering its rededicated capital. When Jewish sources mention Hadrian it is always with the epitaph "may his bones be crushed" (תומצע קיחש or אימט קיחש, the Aramaic equivalent[59]), an expression never used even with respect to Vespasian or Titus who destroyed the Second Temple.

Final years

Succession

Hadrian spent the final years of his life at Rome. In 134, he took an Imperial salutation or the end of the Second Jewish War (which was not actually concluded until the following year). In 136, he dedicated a new Temple of Venus and Roma on the former site of Nero's Golden House.

A bust of Hadrian, National Archaeological Museum of Athens.

About this time, suffering from poor health, he turned to the problem of the succession. In 136 he adopted one of the ordinary consuls of that year, Lucius Ceionius Commodus, who took the name Lucius Aelius Caesar. He was the son-in-law of Gaius Avidius Nigrinus, one of the "four consulars" executed in 118, but was himself in delicate health. Granted tribunician power and the governorship of Pannonia, Aelius Caesar held a further consulship in 137, but died on 1 January 138.[60]

Following the death of Aelius Caesar, Hadrian next adopted Titus Aurelius Fulvus Boionius Arrius Antoninus (the future emperor Antoninus Pius), who had served as one of the four imperial legates of Italy (a post created by Hadrian) and as proconsul of Asia. On 25 February 138 Antoninus received tribunician power and imperium. Moreover, to ensure the future of the dynasty, Hadrian required Antoninus to adopt both Lucius Ceionius Commodus (son of the deceased Aelius Caesar) and Marcus Annius Verus (who was the grandson of an influential senator of the same name who had been Hadrian's close friend; Annius was already betrothed to Aelius Caesar's daughter Ceionia Fabia). Hadrian's precise intentions in this arrangement are debatable.

Though the consensus is that he wanted Annius Verus (who would later become the Emperor Marcus Aurelius) to succeed Antoninus, it has also been argued that he actually intended Ceionius Commodus, the son of his own adopted son, to succeed, but was constrained to show favour simultaneously to Annius Verus because of his strong connections to the Hispano-Narbonensian nexus of senatorial families of which Hadrian himself was a part.[61]

It may well not have been Hadrian, but rather Antoninus Pius — who was Annius Verus's uncle – who advanced the latter to the principal position. The fact that Annius would divorce Ceionia Fabia and re-marry to Antoninus' daughter Annia Faustina points in the same direction. When he eventually became Emperor, Marcus Aurelius would co-opt Ceionius Commodus as his co-Emperor (under the name of Lucius Verus) on his own initiative.[61]

The ancient sources present Hadrian's last few years as marked by conflict and unhappiness. The adoption of Aelius Caesar proved unpopular, not least with Hadrian's brother-in-law Lucius Julius Ursus Servianus and Servianus' grandson Gnaeus Pedanius Fuscus Salinator. Servianus, though now far too old, had stood in line of succession at the beginning of the reign; Fuscus is said to have had designs on the imperial power for himself, and in 137 he may have attempted a coup in which his grandfather was implicated. Whatever the truth, Hadrian ordered that both be put to death.[62] Servianus is reported to have prayed before his execution that Hadrian would "long for death but be unable to die".[63] The prayer was fulfilled; as Hadrian suffered from his final, protracted illness, he had to be prevented from suicide on several occasions.[64]

Death

Hadrian died in the year 138 on the 10th of July, in his villa at Baiae at the age of 62. The cause of death is believed to have been heart failure. Dio Cassius and the Historia Augusta record details of his failing health.

He was buried first at Puteoli, near Baiae, on an estate which had once belonged to Cicero. Soon after, his remains were transferred to Rome and buried in the Gardens of Domitia, close by the almost-complete mausoleum. Upon completion of the Tomb of Hadrian in Rome in 139 by his successor Antoninus Pius, his body was cremated, and his ashes were placed there together with those of his wife Vibia Sabina and his first adopted son, Lucius Aelius, who also died in 138. Antoninus also had him deified in 139 and given a temple on the Campus Martius.

Poem by Hadrian

According to the Historia Augusta, Hadrian composed shortly before his death the following poem:[65]

Animula, vagula, blandula

Hospes comesque corporis

Quae nunc abibis in loca

Pallidula, rigida, nudula,

Nec, ut soles, dabis iocos...

P. Aelius Hadrianus Imp.

Roving amiable little soul,

Body's companion and guest,

Now descending for parts

Colourless, unbending, and bare

Your usual distractions no more shall be there...

Nerva–Antonine family tree

- (1) = 1st spouse
- (2) = 2nd spouse (not shown)
- (3) = 3rd spouse
- SMALL CAPS = posthumously deified (*Augusti, Augustae,* or other)
- dotted lines indicate adoption or (in the case of Hadrian and Antinous) alleged lovers

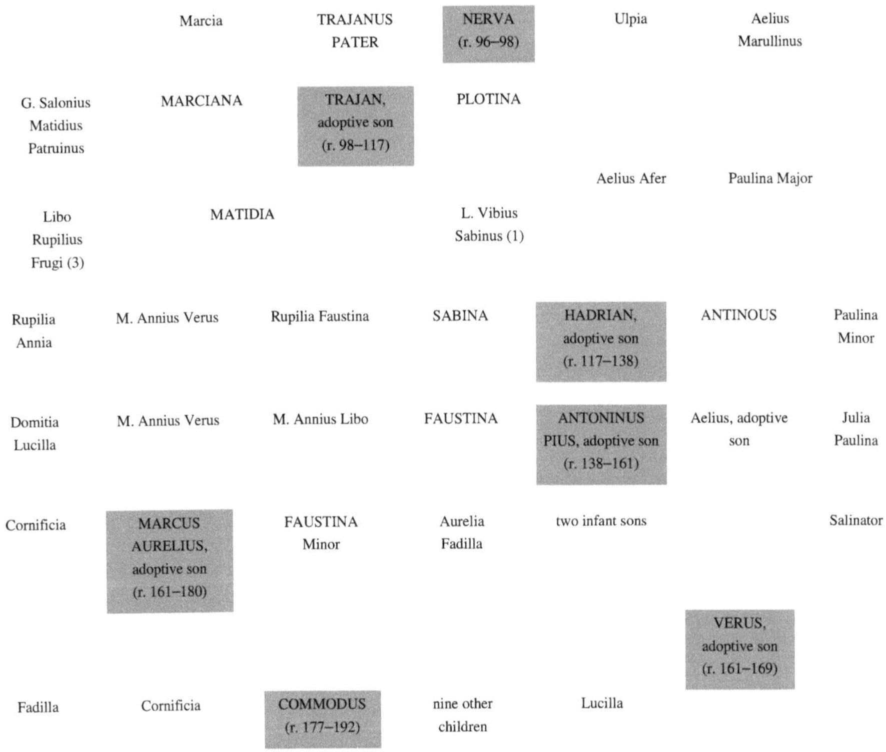

Notes

[1] In Classical Latin, Hadrian's name would be inscribed as PVBLIVS AELIVS TRAIANVS HADRIANVS AVGVSTVS.

[2] Inscription in Athens, year 112 AD: CIL III, 550 = InscrAtt 3 = IG II, 3286 = Dessau 308 = IDRE 2, 365: *P(ublio) Aelio P(ubli) f(ilio) Serg(ia) Hadriano / co(n)s(uli) VIIviro epulonum sodali Augustali leg(ato) pro pr(aetore) Imp(eratoris) Nervae Traiani / Caesaris Aug(usti) Germanici Dacici Pannoniae inferioris praetori eodemque / tempore leg(ato) leg(ionis) I Minerviae P(iae) F(idelis) bello Dacico item trib(uno) pleb(is) quaestori Imperatoris / Traiani et comiti expeditionis Dacicae donis militaribus ab eo donato bis trib(uno) leg(ionis) II / Adiutricis P(iae) F(idelis) item legionis V Macedonicae item legionis XXII Primigeniae P(iae) F(idelis) seviro / turmae eq(uitum) R(omanorum) praef(ecto) feriarum Latinarum Xviro s(tlitibus) i(udicandis) //...*(text in greek)

[3] As emperor his name was *Imperator Caesar Divi Traiani filius Traianus Hadrianus Augustus.*

[4] Eutr. VIII. 6 (http://www.forumromanum.org/literature/eutropius/text8.html): "...nam eum (Hadrianum) Traianus, quamquam consobrinae suae filium..." and SHA, *Vita Hadr.* I, 2 (http://penelope.uchicago.edu/Thayer/L/Roman/Texts/Historia_Augusta/Hadrian/1*.html): *...pater Aelius Hadrianus cognomento Afer fuit, consobrinus Traiani imperatoris.*

[5] After A. M. Canto, in UCM.es (http://www.ucm.es/BUCM/revistas/ghi/02130181/articulos/GERI0303120305A.PDF), specifically pp. 322, 328, 341 and footnote 124, where she stands out SHA, *Vita Hadr.* 1.2: *pro filio habitus* (years 93); 3.2: *ad bellum Dacicum Traianum*

familiarius prosecutus est (year 101) or, principally, 3.7: *quare adamante gemma quam Traianus a Nerva acceperat donatus ad spem successionis erectus est* (year 107).

[6] Royston Lambert, 1984, p. 175

[7] Alicia M. Canto, "Itálica, *patria* y ciudad natal de Adriano (31 textos históricos y argumentos contra *Vita Hadr.* His father died in AD 86 when Hadrian was at the age of 10. 1, 3", *Athenaeum* vol. 92.2, 2004, pp. 367–408 UNIPV.it (http://dobc.unipv.it/dipscant/athenaeum/athenaeum.html)

[8] Ronald Syme, in "Hadrian and Italica" (*Journal of Roman Studies*, LIV, 1964; pp. 142-9) examined the question and concluded that Rome was his birthplace. Against this is the argument of Canto, who argues that only one ancient source gives Hadrian's birth as Rome (SHA, *Vita Hadr* 2,4, probably interpolated), opposite to 25 ancient authors who affirm that he was born in Italica. Among these ancient sources is included his own imperial horoscope, which remained in the famous Antigonus of Nicaea's collection (end of the 2nd. century). This horoscope was well studied by prominent authors as F. H. Cramer, *Astrology in Roman Law and Politics*, Mem.Amer.Philos.Soc. nr. 37 , Philadelphia, 1954 (repr. 1996), see (http://books.google.es/books?id=zv0UAAAAIAAJ&pgis=1) for Hadrian p. 162–178, fn. 121b and 122, etc.: "*...Hadrian – whose horoscope is absolutely certain – surely was born in southern Spain... (in) SHA, Hadrian, 2, 4, the birth was erroneously assigned to Rome instead of Italica, the actual birth-place of Hadrian...*", or O. Neugebauer and H. B. Van Hoesen in their magisterial compilation *Greek Horoscopes*, Mem.Amer.Philos.Soc. nr. 48, Philadelphia, 1959, nr. L76, see now here, ed. 1987 (http://books.google.es/books?id=kEgnLpm06zQC) pp. 80, 90–1, and his footnote 19. They came also to the conclusion that the astronomic parallel of the Hadrian's birth is situated in the Baetica, today Andalusia: "*...L40 agrees exactly with the geographical latitude of southern Spain, the place of origin of Hadrian and his family...*".. "*since Hadrian was born in Italica (southern Spain, near Seville, latitude about 37° 30)...*".

[9] *Historia Augusta*, 'Hadrian', I-II, here explicitly citing the autobiography. This is one of the passages in the Historia Augusta where there is no reason to suspect invention. But see now the Canto's 31 contrary arguments in the *op.cit. supra*; among them, in the same Historia Augusta and, from the same author, Aelius Spartianus, *Vita Sev.* 21: *Falsus est etiam ipse Traianus in* ***suo municipe*** *ac nepote diligendo*, see also es:Adriano#cite note-nacimiento-0, and, characterizing him as a man of provinces (Canto, *ibid.*): *Vita Hadr.* 1,3: *Quaesturam gessit Traiano quater et Articuleio consulibus, in qua cum orationem imperatoris in senatu* ***agrestius pronuntians risus esset****, usque ad summam peritiam et facundiam Latinis operam dedit*

[10] On the numerous senatorial families from Spain residing at Rome and its vicinity around the time of Hadrian's birth see R.Syme, 'Spaniards at Tivoli', in *Roman Papers IV* (Oxford, 1988), pp. 96–114. Tivoli (Tibur) was of course the site of Hadrian's own imperial villa

[11] Royston Lambert, *Beloved And God*, pp.31–32.

[12] Aul.Gell., *Noct.Att.* XVI, 13, 4, and some inscriptions in the city with *C(olonia) A(elia) A(ugusta) I(talica)*

[13] The inscription in footnote 1

[14] H. W. Benario in Roman-emperors.org (http://www.roman-emperors.org/hadrian.htm)

[15] Anthony Birley, *Hadrian the Restless Emperor*, p. 68

[16] Anthony Birley, p. 75

[17] Elizabeth Speller, p. 25

[18] Royston Lambert, p. 34

[19] Elizabeth Speller.

[20] Elizabeth Speller, p. 69

[21] Lambert, Royston, *Beloved and God*, New York: Viking, 1984, pp. 1–14 *passim.*

[22] Edward Gibbon, The Decline and Fall of the Roman Empire, Modern Library Edition, V. 1, p. 68, n. 49.

[23] *Historia Augusta* (c. 395) Hadr. 14.5–7

[24] "Antinous's mysterious death in the Nile led to a Graeco-Egyptian hero-cult to surpass all others in the Greek-speaking world, and busts of the young man are now among the most common from antiquity." (MacGregor, Neil, " There's more to Hadrian than wall-building (http://entertainment.timesonline.co.uk/tol/arts_and_entertainment/visual_arts/article4353046.ece)", *Times" of London, 6 July 2008.*
Dyson, Stephen L., Rome: A Living Portrait of an Ancient City, *p. 195.*

[25] "The public taking of Antinous the Greek as a lover makes more sense as a deliberate political manoeuvre designed to ingratiate himself with the Greek-speakers who still made up 50% of the empire." (Januszczak, Waldemar, " Hadrian – Empire and Conflict at the British Museum (http://entertainment.timesonline.co.uk/tol/arts_and_entertainment/visual_arts/article4353046.ece)", *Times* of London, 20 July 2008)
Lambert, *op. cit.* p. 185.

[26] *Historia Augusta, Hadrian* 2.1 (http://penelope.uchicago.edu/Thayer/E/Roman/Texts/Historia_Augusta/Hadrian/1*.html#2).

[27] Fox, Robin *The Classical World: An Epic History from Homer to Hadrian* Basic Books. 2006 pg 574

[28] Conway, A. E. (1914). "?". *The Burlington Magazine for Connoisseurs* **25** (138): 346–349. JSTOR 859783.

[29] "Facts About the Byzantine Emperors" (http://web2.airmail.net/uthman/byzantine.html). Web2.airmail.net. 2001-09-07. . Retrieved 2012-05-07.

[30] Elizabeth Speller, pp. 74–81

[31] The *Historia Augusta* notes that 'the Britons could not be kept under Roman control, Pompeius Falco was sent to Britain to restore order (Birley 123) and coins of 119–120 refer to this.

[32] *Scriptores Historiae Augustae*, Hadrian, xi, 2

[33] Breeze, David J., and Brian Dobson, "Hadrian's Wall: Some Problems", *Britannia*, Vol. 3, (1972), pp. 182–208

[34] Birley, pp. 131–3

[35] Breeze and Dobson (2000) pp. 15–7
[36] "Britannia on British Coins" (http://www.24carat.co.uk/britanniaframe.html). Chard. . Retrieved 2006-06-25.
[37] Royston Lambert, pp. 41–2
[38] Anthony Birley, pp. 151–2
[39] Anthony Birley, pp. 153–5
[40] Anthony Birley, pp. 157–8
[41] Royston Lambert, pp. 60–1
[42] Anthony Birley, pp. 164–7
[43] Anthony Birley, pp. 175–7
[44] Anthony Birley, pp. 177–80
[45] Anthony Birley, pp. 182–4
[46] Anthony Birley, pp. 189–190
[47] Anthony Birley, pp. 191–200
[48] Royston Lambert, pp. 71–2
[49] Anthony Birley, pp. 213–4
[50] Anthony Birley, pp. 215–20
[51] Cassius Dio, LIX.11; *Historia Augusta, Hadrian*
[52] Ronald Syme, "Journeys of Hadrian" (1988), pp. 164–9
[53] Virgilio Corbo, *The Holy Sepulchre of Jerusalem* (1981)
[54] Cassius Dio, Roman history 69.12.1
[55] Historia Augusta, *Hadrian* 14.2 (http://penelope.uchicago.edu/Thayer/E/Roman/Texts/Historia_Augusta/Hadrian/1*.html#14.2)
[56] livius.org account (http://www.livius.org/le-lh/legio/xxii_deiotariana.html)(Legio XXII Deiotariana)
[57] Cassius Dio 69, 14.3
[58] Gittin 57a-58b; Lamentations Rabbah 2.2 §4;
[59] The Aramaic version, "אימט קיחש", is used, e.g., in Genesis Rabbah 78:1 (http://he.wikisource.org/wiki/×××¨×©_×¨××_×¢×_×××©××). This is referenced by Rashi in his comment on the phrase, "שפנל אמט", in his commentary on Numbers 5:2 (http://www.chabad.org/parshah/torahreading.asp?AID=39589&p=3&showrashi=true). The other two locations in Genesis Rabbah referenced in Rashi's comment, 10:3 and 28:3, use the Hebrew version, "תומצע קיחש"
[60] Anthony Birley, pp. 289–292.
[61] The adoptions: Anthony Birley, pp. 294–5; T.D. Barnes, 'Hadrian and Lucius Verus', *Journal of Roman Studies* (1967), Ronald Syme, *Tacitus*, p. 601. Antoninus as a legate of Italy: Anthony Birley, p. 199
[62] Anthony Birley, pp. 291–2
[63] Dio 69.17.2 (http://penelope.uchicago.edu/Thayer/E/Roman/Texts/Cassius_Dio/69*.html#17.2)
[64] Anthony Birley, p. 297
[65] Historia Augusta, *Hadrian* 25.9 (http://penelope.uchicago.edu/Thayer/E/Roman/Texts/Historia_Augusta/Hadrian/2*.html#25.9); Antony Birley, p. 301

References

Primary sources

- Cassius Dio or Dio Cassius *Roman History*. Greek Text and Translation by Earnest Cary (http://www.archive.org/stream/diosromanhistory08cassuoft#page/424/mode/2up) at internet archive
- Scriptores Historiae Augustae, *Augustan History*. Latin Text (http://penelope.uchicago.edu/Thayer/E/Roman/Texts/Historia_Augusta/Hadrian/1*.html) Translated by David Magie (http://penelope.uchicago.edu/Thayer/E/Roman/Texts/Historia_Augusta/Hadrian/1*.html)
- Aurelius Victor, *Caesares*, XIV. Latin "Caesares: text – IntraText CT" (http://www.intratext.com/IXT/LAT0199/__P5.HTM). Intratext.com. 2007-05-04. Retrieved 2010-03-13.
- Anon, *Excerpta* of Aurelius Victor: *Epitome de Caesaribus*, XIII. Latin "Epitome De Caesaribus: text – IntraText CT" (http://www.intratext.com/IXT/LAT0210/__P5.HTM). Intratext.com. 2007-05-04. Retrieved 2010-03-13.

Inscriptions:

- Eusebius of Caesarea, *Church History (Book IV)*, "Church History" (http://www.newadvent.org/fathers/250104.htm). www.newadvent.org. Retrieved 2010-03-13.
- Smallwood, E.M, *Documents Illustrating the Principates of Nerva Trajan and Hadrian*, Cambridge, 1966.

Secondary sources

- Barnes, T. D. (1967). "Hadrian and Lucius Verus". *Journal of Roman Studies* **57** (1/2): 65–79. doi:10.2307/299345. JSTOR 299345.
- Birley, Anthony R. (1997). *Hadrian. The restless emperor*. London: Routledge. ISBN 0-415-16544-X.
- Canto, Alicia M. (2004). "Itálica, *patria* y ciudad natal de Adriano (31 textos históricos y argumentos contra *Vita Hadr*. 1, 3" (http://dobc.unipv.it/dipscant/athenaeum/athenaeum.html). *Athenaeum* **92.2**: 367–408.
- Dobson, Brian (2000). *Hadrian's Wall*. London: Penguin.
- Gibbon, Edward, *The Decline and Fall of the Roman Empire*, vol. I, 1776. The Online Library of Liberty "Online Library of Liberty – The History of the Decline and Fall of the Roman Empire, vol. 1" (http://oll.libertyfund.org/index.php?option=com_staticxt&staticfile=show.php?title=1365&Itemid=27). Oll.libertyfund.org. Retrieved 2010-03-13.
- Lambert, Royston (1997). *Beloved and God: the story of Hadrian and Antinous*. London: Phoenix Giants. ISBN 1-85799-944-4.
- Speller, Elizabeth (2003). *Following Hadrian: a second-century journey through the Roman Empire*. London: Review. ISBN 0-7472-6662-X.
- Syme, Ronald (1997) [1958]. *Tacitus*. Oxford: Oxford University Press. ISBN 0-19-814327-3.
- Syme, Ronald (1964). "Hadrian and Italica". *Journal of Roman Studies* **LIV**: 142–9.
- Syme, Ronald (1988). "Journeys of Hadrian" (http://www.uni-koeln.de/phil-fak/ifa/zpe/downloads/1988/073pdf/073159.pdf) (PDF). *Zeitschrift für Papyrologie und Epigraphik* **73**: 159–170. Retrieved 2006-12-12. Reprinted in Syme, Ronald (1991). *Roman Papers VI*. Oxford: Clarendon Press. pp. 346–357. ISBN 0-19-814494-6.
- Yourcenar, Marguerite (2005) [1951]. *Memoirs of Hadrian*. New York: Farrar, Straus and Giroux. ISBN 0-374-52926-4.

Further reading

- Boatwright, Mary T. (2003). *Hadrian and the cities of the roman empire.*. Princeton: Princeton Univ Press. ISBN 0-691-09493-4.
- Danziger, Danny; Purcell, Nicholas (2006). *Hadrian's empire : when Rome ruled the world*. London: Hodder & Stoughton. ISBN 0-340-83361-0.
- Everitt, Anthony (2009). *Hadrian and the triumph of Rome*. New York: Random House. ISBN 978-1-4000-6662-9.
- Gray, William Dodge (1919). "A Study of the life of Hadrian Prior to His Accession". *Smith College Studies in History* **4**: 151–209.
- Gregorovius, Ferdinand (1898). *The Emperor Hadrian: A Picture of the Greco-Roman World in His Time* (http://books.google.com/books?id=D20G5zMpPfUC). Mary E. Robinson, trans. London: Macmillan.
- Henderson, Bernard W. (1923). *Life and Principate of the Emperor Hadrian*. London: Methuen.
- Ish-Kishor, Sulamith (1935). *Magnificent Hadrian: A Biography of Hadrian, Emperor of Rome*. New York: Minton, Balch and Co.
- Perowne, Stewart (1960). *Hadrian*. London: Hodder and Stoughton.

External links

- Historia Augusta: Life of Hadrian (http://penelope.uchicago.edu/Thayer/E/Roman/Texts/Historia_Augusta/Hadrian/1*.html)
- Hadrian coinage (http://www.wildwinds.com/coins/ric/hadrian/t.html)
- Temple of Hadrian (http://www.britishtours.com/rome/piazza_di_pietra.html) QuickTime VR, Rome
- Catholic Encyclopedia article (http://www.newadvent.org/cathen/07104b.htm)
- A Bibliography (http://bsa.biblio.univ-lille3.fr/hadrien.htm)
- Major scultoric find at Sagalassos (Turkey) (http://www.archaeology.org/online/features/hadrian/), 2 August 2007 (between 13 and 16 feet in height, four to five meters), with some splendid photos courtesy of the Sagalassos Archaeological Research Project (http://www.archaeology.org/online/features/hadrian/1.html)
- Hadrian, in: De Imperatoribus Romanis, An Online Encyclopedia of Roman Emperors (http://www.roman-emperors.org/hadrian.htm)
- Next exhibition on Hadrian in the British Museum (http://www.britishmuseum.org/whats_on/future_exhibitions/hadrian.aspx), 24 July – 26 October 2008: "Hadrian, Empire and Conflict". Curator: Thorsten Opper
- "Emperor Hadrian, YouTube hero" (http://entertainment.timesonline.co.uk/tol/arts_and_entertainment/the_tls/article4470548.ece): a review by Tom Holland of the Hadrian Exhibition at the British Museum, TLS (http://www.the-tls.co.uk), 6 August 2008.

Conic_section

In mathematics, a **conic section** (or just **conic**) is a curve obtained as the intersection of a cone (more precisely, a right circular conical surface) with a plane. In analytic geometry, a conic may be defined as a plane algebraic curve of degree 2. There are a number of other geometric definitions possible. One of the most useful, in that it involves only the plane, is that a conic consists of those points whose distances to some point, called a *focus*, and some line, called a *directrix*, are in a fixed ratio, called the *eccentricity*.

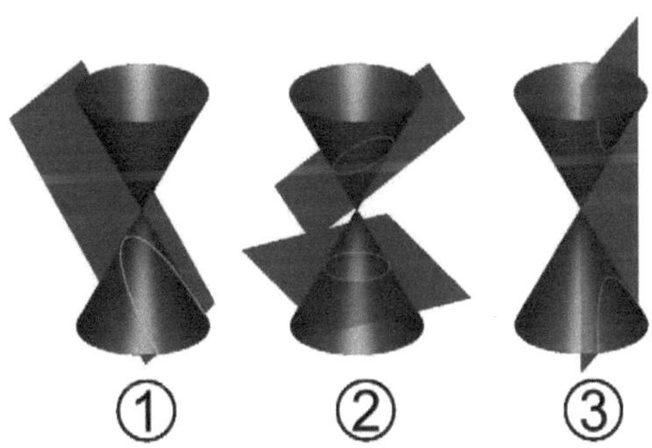

Types of conic sections:
1. Parabola
2. Circle and ellipse
3. Hyperbola

Traditionally, the three types of conic section are the hyperbola, the parabola, and the ellipse. The circle is a special case of the ellipse, and is of sufficient interest in its own right that it is sometimes called the fourth type of conic section. The type of a conic corresponds to its eccentricity, those with eccentricity less than 1 being ellipses, those with eccentricity equal to 1 being

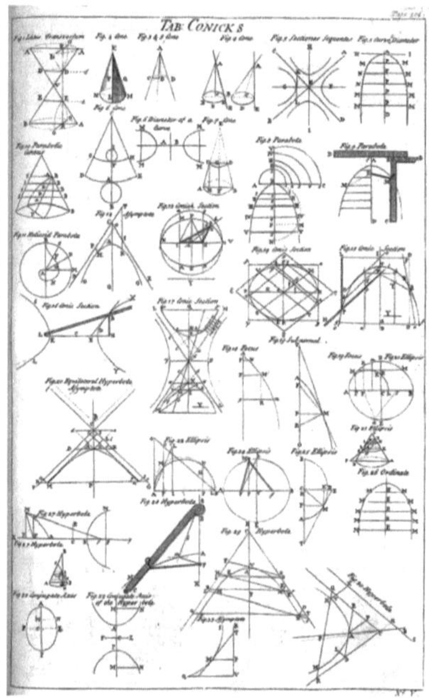

Table of conics, *Cyclopaedia*, 1728

parabolas, and those with eccentricity greater than 1 being hyperbolas. In the focus-directrix definition of a conic the circle is a limiting case with eccentricity 0. In modern geometry certain degenerate cases, such as the union of two lines, are included as conics as well.

The conic sections were named and studied at least since 200 BC, when Apollonius of Perga undertook a systematic study of their properties.

History

Menaechmus and early works

It is believed that the first definition of a conic section is due to Menaechmus (died 320 BC). His work did not survive and is only known through secondary accounts. The definition used at that time differs from the one commonly used today in that it requires the plane cutting the cone to be perpendicular to one of the lines, (a generatrix), that generates the cone as a surface of revolution. Thus the shape of the conic is determined by the angle formed at the vertex of the cone (between two opposite generatrices): If the angle is acute then the conic is an ellipse; if the angle is right then the conic is a parabola; and if the angle is obtuse then the conic is a hyperbola. Note that the circle cannot be defined this way and was not considered a conic at this time.

Euclid (fl. 300 BC) is said to have written four books on conics but these were lost as well.[1] Archimedes (died c. 212 BC) is known to have studied conics, having determined the area bounded by a parabola and an ellipse. The only part of this work to survive is a book on the solids of revolution of conics.

Apollonius of Perga

The greatest progress in the study of conics by the ancient Greeks is due to Apollonius of Perga (died c.190 BC)), whose eight volume *Conic Sections* summarized the existing knowledge at the time and greatly extended it. Apollonius's major innovation was to characterize a conic using properties within the plane and intrinsic to the curve; this greatly simplified analysis. With this tool, it was now possible to show that any plane cutting the cone, regardless of its angle, will produce a conic according to the earlier definition, leading to the definition commonly used today.

Pappus of Alexandria (died c. 350 CE) is credited with discovering importance of the concept of a focus of a conic, and the discovery of the related concept of a directrix.

Al-Kuhi

An instrument for drawing conic sections was first described in 1000 CE by the Islamic mathematician Al-Kuhi.[2][3]

Omar Khayyám

Apollonius's work was translated into Arabic (the technical language of the time) and much of his work only survives through the Arabic version. Persians found applications to the theory; the most notable of these was the Persian[4] mathematician and poet Omar Khayyám who used conic sections to solve algebraic equations.

Europe

Johannes Kepler extended the theory of conics through the "principle of continuity", a precursor to the concept of limits. Girard Desargues and Blaise Pascal developed a theory of conics using an early form of projective geometry and this helped to provide impetus for the study of this new field. In particular, Pascal discovered a theorem known as the hexagrammum mysticum from which many other properties of conics can be deduced. Meanwhile, René Descartes applied his newly discovered Analytic geometry to the study of conics. This had the effect of reducing the geometrical problems of conics to problems in algebra.

Features

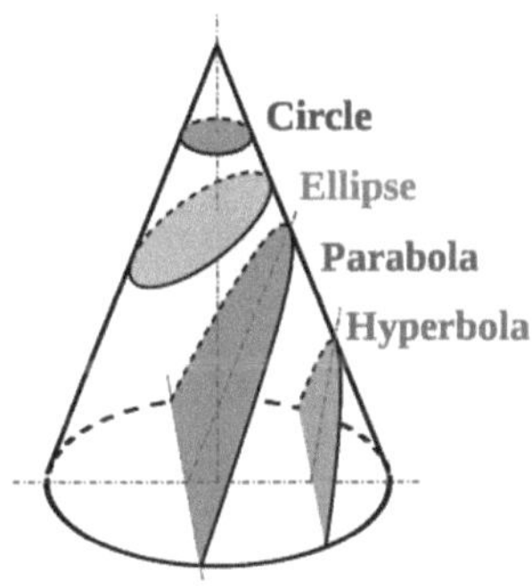

Conics are of three types: parabolas, ellipses, including circles, and hyperbolas.

The three types of conics are the ellipse, parabola, and hyperbola. The circle can be considered as a fourth type (as it was by Apollonius) or as a kind of ellipse. The circle and the ellipse arise when the intersection of cone and plane is a closed curve. The circle is obtained when the cutting plane is parallel to the plane of the generating circle of the cone – for a right cone as in the picture at the top of the page this means that the cutting plane is perpendicular to the symmetry axis of the cone. If the cutting plane is parallel to exactly one generating line of the cone, then the conic is unbounded and is called a parabola. In the remaining case, the figure is a hyperbola. In this case, the plane will intersect *both* halves (*nappes*) of the cone, producing two separate unbounded curves.

Various parameters are associated with a conic section, as shown in the following table. (For the ellipse, the table gives the case of $a>b$, for which the major axis is horizontal; for the reverse case, interchange the symbols a and b. For the hyperbola the east-west opening case is given. In all cases, a and b are positive.)

conic section	equation	eccentricity (e)	linear eccentricity (c)	semi-latus rectum ()	focal parameter (p)
circle					
ellipse					
parabola					
hyperbola					

Conic sections are exactly those curves that, for a point F, a line L not containing F and a non-negative number e, are the locus of points whose distance to F equals e times their distance to L. F is called the focus, L the directrix, and e the **eccentricity**.

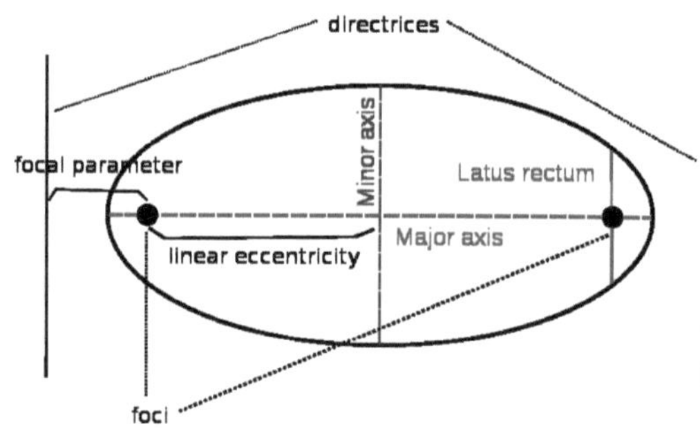

Conic parameters in the case of an ellipse

The **linear eccentricity** (c) is the distance between the center and the focus (or one of the two foci).

The **latus rectum** (2) is the chord parallel to the directrix and passing through the focus (or one of the two foci).

The **semi-latus rectum** () is half the latus rectum.

The **focal parameter** (p) is the distance from the focus (or one of the two foci) to the directrix.

The following relations hold:

Construction

There are many methods to construct a conic. One of them, that is useful in engineering applications, being parallelogram method, where a conic is constructed point by point by means of connecting certain equally spaced points on horizontal line and vertical line.

Properties

Just as two (distinct) points determine a line, five points determine a conic. Formally, given any five points in the plane in general linear position, meaning no three collinear, there is a unique conic passing through them, which will be non-degenerate; this is true over both the affine plane and projective plane. Indeed, given any five points there is a conic passing through them, but if three of the points are collinear the conic will be degenerate (reducible, because it contains a line), and may not be unique; see further discussion.

Irreducible conic sections are always "smooth". This is important for many applications, such as aerodynamics, where a smooth surface is required to ensure laminar flow and to prevent turbulence.

Intersection at infinity

An algebro-geometrically intrinsic form of this classification is by the intersection of the conic with the line at infinity, which gives further insight into their geometry:

- ellipses intersect the line at infinity in 0 points – rather, in 0 real points, but in 2 complex points, which are conjugate;
- parabolas intersect the line at infinity in 1 double point, corresponding to the axis – they are tangent to the line at infinity, and close at infinity, as distended ellipses;
- hyperbolas intersect the line at infinity in 2 points, corresponding to the asymptotes – hyperbolas pass through infinity, with a twist. Going to infinity along one branch passes through the point at infinity corresponding to the asymptote, then re-emerges on the other branch at the other side but with the inside of the hyperbola (the direction of curvature) on the other side – left vs. right (corresponding to the non-orientability of the real projective plane) – and then passing through the other point at infinity returns to the first branch. Hyperbolas can thus be seen as ellipses that have been pulled through infinity and re-emerged on the other side, flipped.

Degenerate cases

There are five degenerate cases: three in which the plane passes through apex of the cone, and three that arise when the cone itself degenerates to a cylinder (a doubled line can occur in both cases).

When the plane passes through the apex, the resulting conic is always degenerate, and is either: a point (when the angle between the plane and the axis of the cone is larger than tangential); a straight line (when the plane is tangential to the surface of the cone); or a pair of intersecting lines (when the angle is smaller than the tangential). These correspond respectively to degeneration of an ellipse, parabola, and a hyperbola, which are characterized in the same way by angle. The straight line is more precisely a *double* line (a line with multiplicity 2) because the plane is tangent to the cone, and thus the intersection should be counted twice.

Where the cone is a cylinder, i.e. with the vertex at infinity, cylindric sections are obtained;[5] this corresponds to the apex being at infinity. Cylindrical sections are ellipses (or circles), unless the plane is vertical (which corresponds to passing through the apex at infinity), in which case three degenerate cases occur: two parallel lines, known as a ribbon (corresponding to an ellipse with one axis infinite and the other axis real and non-zero, the distance between the lines), a double line (an ellipse with one infinite axis and one axis zero), and no intersection (an ellipse with one infinite axis and the other axis imaginary).

Eccentricity, focus and directrix

The four defining conditions above can be combined into one condition that depends on a fixed point (the ***focus***), a line (the ***directrix***) not containing and a nonnegative real number (the *eccentricity*). The corresponding conic section consists of the locus of all points whose distance to equals times their distance to . For we obtain an ellipse, for a parabola, and for a hyperbola.

For an ellipse and a hyperbola, two focus-directrix combinations can be taken, each giving the same full ellipse or hyperbola. The distance from the center to the directrix is , where is the semi-major axis of the ellipse, or the distance from the center to the tops of the hyperbola. The distance from the center to a focus is .

In the case of a circle, the eccentricity , and one can imagine the directrix to be infinitely far removed from the center. However, the statement that the circle consists of all points whose distance to F is e times the distance to L is not useful, because we get zero times infinity.

The eccentricity of a conic section is thus a measure of how far it deviates from being circular.

For a given , the closer is to 1, the smaller is the semi-minor axis.

Generalizations

Conics may be defined over other fields, and may also be classified in the projective plane rather than in the affine plane.

Over the complex numbers ellipses and hyperbolas are not distinct, since there is no meaningful difference between 1 and −1; precisely, the ellipse becomes a hyperbola under the substitution geometrically a complex rotation, yielding – a hyperbola is simply an ellipse with an imaginary axis length. Thus there is a 2-way classification: ellipse/hyperbola and parabola. Geometrically, this corresponds to intersecting the line at infinity in either 2 distinct points (corresponding to two asymptotes) or in 1 double point (corresponding to the axis of a parabola), and thus the real hyperbola is a more suggestive image for the complex ellipse/hyperbola, as it also has 2 (real) intersections with the line at infinity.

In projective space, over any division ring, but in particular over either the real or complex numbers, all non-degenerate conics are equivalent, and thus in projective geometry one simply speaks of "a conic" without specifying a type, as type is not meaningful. Geometrically, the line at infinity is no longer special (distinguished), so while some conics intersect the line at infinity differently, this can be changed by a projective transformation – pulling an ellipse out to infinity or pushing a parabola off infinity to an ellipse or a hyperbola.

In other areas of mathematics

The classification into elliptic, parabolic, and hyperbolic is pervasive in mathematics, and often divides a field into sharply distinct subfields. The classification mostly arises due to the presence of a quadratic form (in two variables this corresponds to the associated discriminant), but can also correspond to eccentricity.

Quadratic form classifications:

quadratic forms

> Quadratic forms over the reals are classified by Sylvester's law of inertia, namely by their positive index, zero index, and negative index: a quadratic form in n variables can be converted to a diagonal form, as where the number of +1 coefficients, k, is the positive index, the number of −1 coefficients, l, is the negative index, and the remaining variables are the zero index m, so In two variables the non-zero quadratic forms are classified as:

- – positive-definite (the negative is also included), corresponding to ellipses,
- – degenerate, corresponding to parabolas, and
- – indefinite, corresponding to hyperbolas.

> In two variables quadratic forms are classified by discriminant, analogously to conics, but in higher dimensions the more useful classification is as *definite,* (all positive or all negative), *degenerate,* (some zeros), or *indefinite* (mix of positive and negative but no zeros). This classification underlies many that follow.

curvature

> The Gaussian curvature of a surface describes the infinitesimal geometry, and may at each point be either positive – elliptic geometry, zero – Euclidean geometry (flat, parabola), or negative – hyperbolic geometry; infinitesimally, to second order the surface looks like the graph of , (or 0), or . Indeed, by the uniformization theorem every surface can be taken to be globally (at every point) positively curved, flat, or negatively curved. In higher dimensions the Riemann curvature tensor is a more complicated object, but manifolds with constant sectional curvature are interesting objects of study, and have strikingly different properties, as discussed at sectional curvature.

Second order PDEs

> Partial differential equations (PDEs) of second order are classified at each point as elliptic, parabolic, or hyperbolic, accordingly as their second order terms correspond to an elliptic, parabolic, or hyperbolic

quadratic form. The behavior and theory of these different types of PDEs are strikingly different – representative examples is that the Poisson equation is elliptic, the heat equation is parabolic, and the wave equation is hyperbolic.

Eccentricity classifications include:

Möbius transformations

Real Möbius transformations (elements of $PSL_2(\mathbf{R})$ or its 2-fold cover, $SL_2(\mathbf{R})$) are classified as elliptic, parabolic, or hyperbolic accordingly as their half-trace is or mirroring the classification by eccentricity.

Variance-to-mean ratio

The variance-to-mean ratio classifies several important families of discrete probability distributions: the constant distribution as circular (eccentricity 0), binomial distributions as elliptical, Poisson distributions as parabolic, and negative binomial distributions as hyperbolic. This is elaborated at cumulants of some discrete probability distributions.

Cartesian coordinates

In the Cartesian coordinate system, the graph of a quadratic equation in two variables is always a conic section – though it may be degenerate, and all conic sections arise in this way. The equation will be of the form

As scaling all six constants yields the same locus of zeros, one can consider conics as points in the five-dimensional projective space

Discriminant classification

The conic sections described by this equation can be classified with the discriminant[6]

If the conic is non-degenerate, then:

- if , the equation represents an ellipse;
  - if and , the equation represents a circle, which is a special case of an ellipse;
- if , the equation represents a parabola;
- if , the equation represents a hyperbola;
  - if we also have , the equation represents a rectangular hyperbola.

To distinguish the degenerate cases from the non-degenerate cases, let Δ be the determinant of the 3×3 matrix [A, $B/2$, $D/2$; $B/2$, C, $E/2$; $D/2$, $E/2$, F]: that is, $\Delta = (AC - B^2/4)F + BED/4 - CD^2/4 - AE^2/4$. Then the conic section is non-degenerate if and only if $\Delta \neq 0$. If $\Delta=0$ we have a point ellipse, two parallel lines (possibly coinciding with each other) in the case of a parabola, or two intersecting lines in the case of a hyperbola.[7] :p.63

Moreover, in the case of a non-degenerate ellipse (with and $\Delta\neq 0$), we have a real ellipse if $C\Delta < 0$ but an imaginary ellipse if $C\Delta > 0$. An example is , which has no real-valued solutions.

Note that A and B are polynomial coefficients, not the lengths of semi-major/minor axis as defined in some sources.

Matrix notation

The above equation can be written in matrix notation as

The type of conic section is solely determined by the determinant of middle matrix: if it is positive, zero, or negative then the conic is an ellipse, parabola, or hyperbola respectively (see geometric meaning of a quadratic form). If both the eigenvalues of the middle matrix are non-zero (i.e. it is an ellipse or a hyperbola), we can do a transformation of variables to obtain

where a,c, and G satisfy and .

The quadratic can also be written as

If the determinant of this 3×3 matrix is non-zero, the conic section is not degenerate. If the determinant equals zero, the conic is a degenerate parabola (two parallel or coinciding lines), a degenerate ellipse (a point ellipse), or a degenerate hyperbola (two intersecting lines).

Note that in the centered equation with constant term *G*, *G* equals minus one times the ratio of the 3×3 determinant to the 2×2 determinant.

As slice of quadratic form

The equation

can be rearranged by taking the affine linear part to the other side, yielding

In this form, a conic section is realized exactly as the intersection of the graph of the quadratic form and the plane Parabolas and hyperbolas can be realized by a horizontal plane (), while ellipses require that the plane be slanted. Degenerate conics correspond to degenerate intersections, such as taking slices such as of a positive-definite form.

Eccentricity in terms of parameters of the quadratic form

When the conic section is written algebraically as

the eccentricity can be written as a function of the parameters of the quadratic equation.[8] If $4AC = B^2$ the conic is a parabola and its eccentricity equals 1 (if it is non-degenerate). Otherwise, assuming the equation represents either a non-degenerate hyperbola or a non-degenerate, non-imaginary ellipse, the eccentricity is given by

where $\eta = 1$ if the determinant of the 3×3 matrix is negative and $\eta = -1$ if that determinant is positive.

Standard form

Through change of coordinates these equations can be put in standard forms:

- Circle:
- Ellipse:
- Parabola:
- Hyperbola:
- Rectangular hyperbola:

Such forms will be symmetrical about the *x*-axis and for the circle, ellipse and hyperbola symmetrical about the *y*-axis.
The rectangular hyperbola however is only symmetrical about the lines and . Therefore its inverse function is exactly the same as its original function.

These standard forms can be written as parametric equations,

- Circle: ,
- Ellipse: ,
- Parabola: ,
- Hyperbola: or .
- Rectangular hyperbola:

Invariants of conics

The trace and determinant of are both invariant with respect to both rotation of axes and translation of the plane (movement of the origin).[9]

The constant term F is invariant under rotation only.

Modified form

For some practical applications, it is important to re-arrange the standard form so that the focal-point can be placed at the origin. The mathematical formulation for a general conic section is then given in the polar form by

and in the Cartesian form by

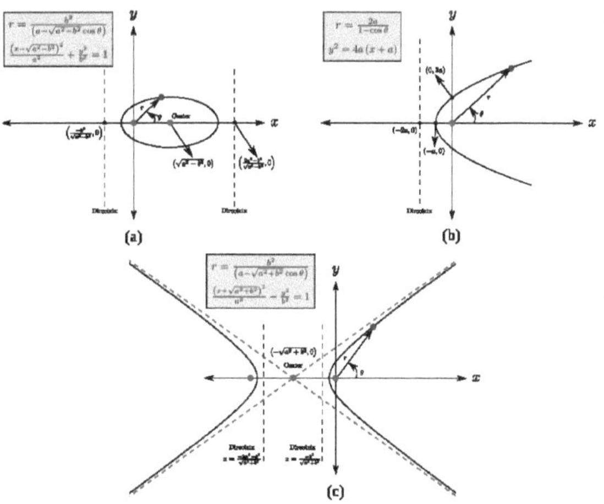

Three different types of conic sections. Focal-points corresponding to all conic sections are placed at the origin.

From the above equation, the **linear eccentricity** (c) is given by .

From the general equations given above, different conic sections can be represented as shown below:

- Circle:
- Ellipse:
- Parabola:
- Hyperbola:

Homogeneous coordinates

In homogeneous coordinates a conic section can be represented as:

Or in matrix notation

The matrix is called *the matrix of the conic section*.

is called the determinant of the conic section. If $\Delta = 0$ then the *conic section* is said to be *degenerate*; this means that the conic section is either a union of two straight lines, a repeated line, a point or the empty set.

For example, the conic section reduces to the union of two lines:

Similarly, a conic section sometimes reduces to a (single) repeated line:

is called the discriminant of the conic section. If $\delta = 0$ then the *conic section* is a parabola, if $\delta < 0$, it is an hyperbola and if $\delta > 0$, it is an ellipse. A conic section is a circle if $\delta > 0$ and $A_1 = A_2$ and $B_1 = 0$, it is an rectangular hyperbola if $\delta < 0$ and $A_1 = -A_2$. It can be proven that in the complex projective plane $\mathbf{CP}^2$ two conic sections have four points in common (if one accounts for multiplicity), so there are never more than 4 intersection points and there is always one *intersection point* (possibilities: four distinct intersection points, two singular intersection points and one double intersection points, two double intersection points, one singular intersection point and 1 with multiplicity 3, 1 intersection point with multiplicity 4). If there exists at least one intersection point with multiplicity > 1, then the two conic sections are said to be tangent. If there is only one intersection point, which has multiplicity 4, the two conic sections are said to be osculating.[10]

Furthermore each straight line intersects each conic section twice. If the intersection point is double, the line is said to be tangent and it is called the tangent line. Because every straight line intersects a conic section twice, each conic section has two points at infinity (the intersection points with the line at infinity). If these points are real, the conic section must be a hyperbola, if they are imaginary conjugated, the conic section must be an ellipse, if the conic section has one double point at infinity it is a parabola. If the points at infinity are (1,i,0) and (1,-i,0), the conic section is a circle (see circular points at infinity). If a conic section has one real and one imaginary point at infinity or it has two imaginary points that are not conjugated then it not a real conic section (its coefficients are complex).

Polar coordinates

In polar coordinates, a conic section with one focus at the origin and, if any, the other on the x-axis, is given by the equation

where e is the eccentricity and l is the semi-latus rectum (see above). As above, for $e = 0$, we have a circle, for $0 < e < 1$ we obtain an ellipse, for $e = 1$ a parabola, and for $e > 1$ a hyperbola.

Pencil of conics

A (non-degenerate) conic is completely determined by five points in general position (no three collinear) in a plane and the system of conics which pass through a fixed set of four points (again in a plane and no three collinear) is called a **pencil of conics**.[11] The four common points are called the *base points* of the pencil. Through any point other than a base point, there passes a single conic of the pencil. This concept generalizes a pencil of circles.

In a projective plane defined over an algebraically closed field any two conics meet in four points (counted with multiplicity) and so, determine the pencil of conics based on these four points. Furthermore, the four base points determine three line pairs (degenerate conics through the base points, each line of the pair containing exactly two base points) and so each pencil of conics will contain at most three degenerate conics.[12]

A pencil of conics can represented algebraically in the following way. Let C_1 and C_2 be two distinct conics in a projective plane defined over an algebraically closed field K. For every pair λ, μ of elements of K, not both zero, the expression:

represents a conic in the pencil determined by C_1 and C_2. This symbolic representation can be made concrete with a slight abuse of notation (using the same notation to denote the object as well as the equation defining the object.) Thinking of C_1, say, as a ternary quadratic form, then $C_1 = 0$ is the equation of the "conic C_1". Another concrete realization would be obtained by thinking of C_1 as the 3×3 symmetric matrix which represents it. If C_1 and C_2 have such concrete realizations then every member of the above pencil will as well. Since the setting uses homogeneous coordinates in a projective plane, two concrete representations (either equations or matrices) give the same conic if they differ by a non-zero multiplicative constant.

Intersecting two conics

The solutions to a system of two second degree equations in two variables may be viewed as the coordinates of the points of intersection of two generic conic sections. In particular two conics may possess none, two or four possibly coincident intersection points. An efficient method of locating these solutions exploits the homogeneous matrix representation of conic sections, i.e. a 3x3 symmetric matrix which depends on six parameters.

The procedure to locate the intersection points follows these steps, where the conics are represented by matrices:

- given the two conics and , consider the pencil of conics given by their linear combination
- identify the homogeneous parameters which correspond to the degenerate conic of the pencil. This can be done by imposing the condition that and solving for and . These turn out to be the solutions of a third degree equation.
- given the degenerate conic , identify the two, possibly coincident, lines constituting it.
- intersect each identified line with either one of the two original conics; this step can be done efficiently using the dual conic representation of
- the points of intersection will represent the solutions to the initial equation system.

Applications

Conic sections are important in astronomy: the orbits of two massive objects that interact according to Newton's law of universal gravitation are conic sections if their common center of mass is considered to be at rest. If they are bound together, they will both trace out ellipses; if they are moving apart, they will both follow parabolas or hyperbolas. See two-body problem.

The paraboloid shape of Archeocyathids produces conic sections on rock faces

In projective geometry, the conic sections in the projective plane are equivalent to each other up to projective transformations.

For specific applications of each type of conic section, see the articles circle, ellipse, parabola, and hyperbola.

For certain fossils in paleontology, understanding conic sections can help understand the three-dimensional shape of certain organisms.

See also

- Focus (geometry), an overview of properties of conic sections related to the foci
- Lambert conformal conic projection
- Matrix representation of conic sections
- Quadrics, the higher-dimensional analogs of conics
- Quadratic function
- Rotation of axes
- Dandelin spheres
- Projective conics
- Elliptic coordinates
- Parabolic coordinates
- Director circle
- Conic Sections Rebellion

Notes

[1] Heath, T.L., *The Thirteen Books of Euclid's Elements*, Vol. I, Dover, 1956, pg.16

[2] Stillwell, John (2010). *Mathematics and its history* (3rd ed. ed.). New York: Springer. p. 30. ISBN 1-4419-6052-X.

[3] "Apollonius of Perga Conics Books One to Seven" (http://www.math.psu.edu/katok_s/Commentaries-new.pdf). . Retrieved 10 June 2011.

[4] Turner, Howard R. (1997). *Science in medieval Islam: an illustrated introduction* (http://books.google.com/books?id=3VfY8PgmhDMC). University of Texas Press. p. 53. ISBN 0-292-78149-0. ., Chapter , p. 53 (http://books.google.com/books?id=3VfY8PgmhDMC&pg=PA53)

[5] "MathWorld: Cylindric section" (http://mathworld.wolfram.com/CylindricSection.html). .

[6] Fanchi, John R. (2006), *Math refresher for scientists and engineers* (http://books.google.com/books?id=75mAJPcAWT8C), John Wiley and Sons, pp. 44–45, ISBN 0-471-75715-2, , Section 3.2, page 45 (http://books.google.com/books?id=75mAJPcAWT8C&pg=PA45)

[7] Lawrence, J. Dennis, *A Catalog of Special Plane Curves*, Dover Publ., 1972.

[8] Ayoub, Ayoub B., "The eccentricity of a conic section," *The College Mathematics Journal* 34(2), March 2003, 116–121.

[9] Pettofrezzo, Anthony, *Matrices and Transformations*, Dover Publ., 1966, pp. 101-111.

[10] Wilczynski, E. J. (1916), "Some remarks on the historical development and the future prospects of the differential geometry of plane curves", *Bull. Amer. Math. Soc.* **22**: 317–329.

[11] Faulkner 1952, pg. 64

[12] Samuel 1988, pg. 50

References

- Akopyan, A.V. and Zaslavsky, A.A. (2007). *Geometry of Conics*. American Mathematical Society. pp. 134. ISBN 0-8218-4323-0.
- Faulkner, T. E. (1952), *Projective Geometry* (2nd ed.), Edinburgh: Oliver and Boyd
- Samuel, Pierre (1988), *Projective Geometry*, Undergraduate Texts in Mathematics (Readings in Mathematics), New York: Springer-Verlag, ISBN 0-387-96752-4

External links

- Derivations of Conic Sections (http://mathdl.maa.org/convergence/1/?pa=content&sa=viewDocument&nodeId=196&bodyId=60) at Convergence (http://mathdl.maa.org/convergence/1/)
- Conic sections (http://xahlee.org/SpecialPlaneCurves_dir/ConicSections_dir/conicSections.html) at Special plane curves (http://xahlee.org/SpecialPlaneCurves_dir/specialPlaneCurves.html).
- Weisstein, Eric W., " Conic Section (http://mathworld.wolfram.com/ConicSection.html)" from MathWorld.
- Determinants and Conic Section Curves (http://math.fullerton.edu/mathews/n2003/ConicFitMod.html)
- Occurrence of the conics. Conics in nature and elsewhere (http://britton.disted.camosun.bc.ca/jbconics.htm).
- Conics (http://www.mathacademy.com/pr/prime/articles/conics/index.asp). An essay on conics and how they are generated.
- See Conic Sections (http://www.cut-the-knot.org/proofs/conics.shtml) at cut-the-knot (http://www.cut-the-knot.org) for a sharp proof that any finite conic section is an ellipse and Xah Lee (http://xahlee.org/PageTwo_dir/more.html) for a similar treatment of other conics.
- Cone-plane intersection (http://www.mathworks.com/matlabcentral/fileexchange/19631) MATLAB code
- Eight Point Conic (http://math.kennesaw.edu/~mdevilli/eightpointconic.html) at Dynamic Geometry Sketches (http://math.kennesaw.edu/~mdevilli/JavaGSPLinks.htm)
- An interactive Java conics grapher; uses a general second-order implicit equation. (http://www.geogebra.org/en/upload/files/nikenuke/conics04b.html)

Apollonius_of_Perga

Apollonius of Perga [Pergaeus] (Ancient Greek: Ἀπολλώνιος) (ca. 262 BC – ca. 190 BC) was a Greek geometer and astronomer noted for his writings on conic sections. His innovative methodology and terminology, especially in the field of conics, influenced many later scholars including Ptolemy, Francesco Maurolico, Johannes Kepler, Isaac Newton, and René Descartes. It was Apollonius who gave the ellipse, the parabola, and the hyperbola the names by which we know them. The hypothesis of eccentric orbits, or equivalently, deferent and epicycles, to explain the apparent motion of the planets and the varying speed of the Moon, is also attributed to him. **Apollonius' theorem** demonstrates that the two models are equivalent given the right parameters. Ptolemy describes this theorem in the *Almagest* XII.1. Apollonius also researched the lunar history, for which he is said to have been called Epsilon (ε). The crater Apollonius on the Moon is named in his honor.

Conics

The degree of originality of the ***Conics*** can best be judged from Apollonius's own prefaces. Books i–iv he describes as an "elementary introduction" containing essential principles, while the other books are specialized investigations in particular directions. He then claims that, in Books i–iv, he only works out the generation of the curves and their fundamental properties presented in Book i more fully and generally than did earlier treatises, and that a number of theorems in Book iii and the greater part of Book iv are new. Allusions to predecessor's works, such as Euclid's four *Books on Conics*, show a debt not only to Euclid but also to Conon and Nicoteles.

The generality of Apollonius's treatment is indeed remarkable. He defines the fundamental conic property as the equivalent of the Cartesian equation applied to *oblique* axes—i.e., axes consisting of a diameter and the tangent at its extremity—that are obtained by cutting an oblique circular cone. The way the cone is cut does not matter. He shows that the oblique axes are only a *particular* case after demonstrating that the basic conic property can be expressed in the same form with reference to *any* new diameter and the tangent at its extremity. It is the form of the fundamental property (expressed in terms of the "application of areas") that leads him to give these curves their names: *parabola*, *ellipse*, and *hyperbola*. Thus Books v–vii are clearly original.

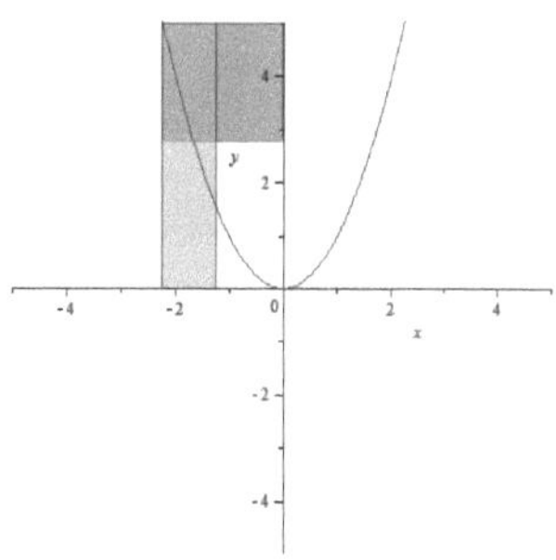

Parabola connection with areas of a square and a rectangle, that inspired Apollonius of Perga to give the parabola its current name.

Apollonius's genius reaches its highest heights in Book v. Here he treats of normals as minimum and maximum straight lines drawn from given points to the curve (independently of tangent properties); discusses how many normals can be drawn from particular points; finds their feet by construction; and gives propositions that both determine the center of curvature at any point and lead at once to the Cartesian equation of the evolute of any conic.

Apollonius in the *Conics* further developed a method that is so similar to analytic geometry that his work is sometimes thought to have anticipated the work of Descartes by some 1800 years. His application of reference lines, a diameter and a tangent is essentially no different than our modern use of a coordinate frame, where the distances measured along the diameter from the point of tangency are the abscissas, and the segments parallel to the tangent and intercepted between the axis and the curve are the ordinates. He further developed relations between the abscissas and the corresponding ordinates that are equivalent to rhetorical equations of curves. However, although Apollonius came close to developing analytic geometry, he did not manage to do so since he did not take into account negative magnitudes and in every case the coordinate system was superimposed upon a given curve *a posteriori* instead of *a priori*. That is, equations were determined by curves, but curves were not determined by equations. Coordinates, variables, and equations were subsidiary notions applied to a specific geometric situation.[1]

Other works

Pappus mentions other treatises of Apollonius:

1. Λόγου ἀποτομή, *De Rationis Sectione* ("Cutting of a Ratio")
2. Χωρίου ἀποτομή, *De Spatii Sectione* ("Cutting of an Area")
3. Διωρισμένη τομή, *De Sectione Determinata* ("Determinate Section")
4. Ἐπαφαί, *De Tactionibus* ("Tangencies")
5. Νεύσεις, *De Inclinationibus* ("Inclinations")
6. Τόποι ἐπίπεδοι, *De Locis Planis* ("Plane Loci").

Each of these was divided into two books, and—with the *Data*, the *Porisms*, and *Surface-Loci* of Euclid and the *Conics* of Apollonius—were, according to Pappus, included in the body of the ancient analysis.

De Rationis Sectione

De Rationis Sectione sought to resolve a simple problem: Given two straight lines and a point in each, draw through a third given point a straight line cutting the two fixed lines such that the parts intercepted between the given points in them and the points of intersection with this third line may have a given ratio.

De Spatii Sectione

De Spatii Sectione discussed a similar problem requiring the rectangle contained by the two intercepts to be equal to a given rectangle.

In the late 17th century, Edward Bernard discovered a version of *De Rationis Sectione* in the Bodleian Library. Although he began a translation, it was Halley who finished it and included it in a 1706 volume with his restoration of *De Spatii Sectione*.

De Sectione Determinata

De Sectione Determinata deals with problems in a manner that may be called an analytic geometry of one dimension; with the question of finding points on a line that were in a ratio to the others.[2] The specific problems are: Given two, three or four points on a straight line, find another point on it such that its distances from the given points satisfy the condition that the square on one or the rectangle contained by two has a given ratio either (1) to the square on the remaining one or the rectangle contained by the remaining two or (2) to the rectangle contained by the remaining one and another given straight line. Several have tried to restore the text to discover Apollonius's solution, among them Snellius (Willebrord Snell, Leiden, 1698); Alexander Anderson of Aberdeen, in the supplement to his *Apollonius Redivivus* (Paris, 1612); and Robert Simson in his *Opera quaedam reliqua* (Glasgow, 1776), by far the best attempt.[3]

De Tactionibus

For more information, see Problem of Apollonius.

De Tactionibus embraced the following general problem: Given three things (points, straight lines, or circles) in position, describe a circle passing through the given points and touching the given straight lines or circles. The most difficult and historically interesting case arises when the three given things are circles. In the 16th century, Vieta presented this problem (sometimes known as the Apollonian Problem) to Adrianus Romanus, who solved it with a hyperbola. Vieta thereupon proposed a simpler solution, eventually leading him to restore the whole of Apollonius's treatise in the small work *Apollonius Gallus* (Paris, 1600). The history of the problem is explored in fascinating detail in the preface to J. W. Camerer's brief *Apollonii Pergaei quae supersunt, ac maxime Lemmata Pappi in hos Libras, cum Observationibus, &c* (Gothae, 1795, 8vo).

De Inclinationibus

The object of *De Inclinationibus* was to demonstrate how a straight line of a given length, tending towards a given point, could be inserted between two given (straight or circular) lines. Though Marin Getaldić and Hugo d'Omerique (*Geometrical Analysis*, Cadiz, 1698) attempted restorations, the best is by Samuel Horsley (1770).[3]

De Locis Planis

De Locis Planis is a collection of propositions relating to loci that are either straight lines or circles. Since Pappus gives somewhat full particulars of its propositions, this text has also seen efforts to restore it, not only by P. Fermat (*Oeuvres*, i., 1891, pp. 3–51) and F. Schooten (Leiden, 1656) but also, most successfully of all, by R. Simson (Glasgow, 1749).[3]

Additional works

Ancient writers refer to other works of Apollonius that are no longer extant:

1. Περὶ τοῦ πυρίου, *On the Burning-Glass*, a treatise probably exploring the focal properties of the parabola
2. Περὶ τοῦ κοχλίου, *On the Cylindrical Helix* (mentioned by Proclus)
3. A comparison of the dodecahedron and the icosahedron inscribed in the same sphere
4. Ἡ καθόλου πραγματεία, a work on the general principles of mathematics that perhaps included Apollonius's criticisms and suggestions for the improvement of Euclid's *Elements*
5. Ὠκυτόκιον ("Quick Bringing-to-birth"), in which, according to Eutocius, Apollonius demonstrated how to find closer limits for the value of π (pi) than those of Archimedes, who calculated 3+1/7 as the upper limit (3.1428571, with the digits after the decimal point repeating) and 3+10/71 as the lower limit (3.1408456338028160, with the digits after the decimal point repeating)
6. an arithmetical work (see Pappus) on a system both for expressing large numbers in language more everyday than that of Archimedes' *The Sand Reckoner* and for multiplying these large numbers
7. a great extension of the theory of irrationals expounded in Euclid, Book x., from binomial to multinomial and from *ordered* to *unordered* irrationals (see extracts from Pappus' comm. on Eucl. x., preserved in Arabic and published by Woepcke, 1856).

Published editions

The best editions of the works of Apollonius are the following:

1. *Apollonii Pergaei Conicorum libri quatuor, ex versione Frederici Commandini* (Bononiae, 1566), fol.
2. *Apollonii Pergaei Conicorum libri octo, et Sereni Antissensis de Sectione Cylindri et Coni libri duo* (Oxoniae, 1710), fol. (this is the monumental edition of Edmund Halley)
3. the edition of the first four books of the Conics given in 1675 by Isaac Barrow
4. *Apollonii Pergaei de Sectione, Rationis libri duo: Accedunt ejusdem de Sectione Spatii libri duo Restituti: Praemittitur, &c., Opera et Studio Edmundi Halley* (Oxoniae, 1706), 4to
5. a German translation of the *Conics* by H. Balsam (Berlin, 1861)
6. The definitive Greek text is the edition of Heiberg (*Apollonii Pergaei quae Graece exstant Opera*, Leipzig, 1891–1893)
7. T. L. Heath, *Apollonius, Treatise on Conic Sections* (Cambridge, 1896)
8. The Arabic translation of the Books V–VII was first published in two volumes by Springer Verlag in 1990 (ISBN 0-387-97216-1), volume 9 in the "Sources in the history of mathematics and physical sciences" series. The edition was produced by G. J. Toomer and provided with an English translation and various commentaries.

9. *Conics: Books I–III* translated by R. Catesby Taliaferro, published by Green Lion Press (ISBN 1-888009-05-5). (An English translation of Book IV by Michael N. Fried is also available from the same publisher. ISBN 1-888009-20-9)
10. *Apollonius de Perge, Coniques: Texte grec et arabe etabli, traduit et commenté* (De Gruyter, 2008–2010), eds. R. Rashed, M. Decorps-Foulquier, M. Federspiel. (This is a new edition of the surviving Greek text (Books I–IV), a full edition of the surviving Arabic text (Books I–VII) with French translation and commentaries.)
11. *Apollonius of Perga's Conica: Text, Context, Subtext.* By Michael N. Fried and Sabetai Unguru (Brill).
12. *Edmund Halley's Reconstruction of the Lost Book of Apollonius' Conics.* By Michael N. Fried (ISBN 1461401453).

See also

- Apollonian circles
- Apollonian gasket
- Apollonian network
- Circles of Apollonius
- Descartes' theorem
- Problem of Apollonius
- Apollonius' theorem

Notes

[1] Boyer, Carl B. (1991). "Apollonius of Perga". *A History of Mathematics* (Second ed.). John Wiley & Sons, Inc.. pp. 156–157. ISBN 0-471-54397-7. "The method of Apollonius in the *Conics* in many respects are so similar to the modern approach that his work sometimes is judged to be an analytic geometry anticipating that of Descartes by 1800 years. The application of references lines in general, and of a diameter and a tangent at its extremity in particular, is, of course, not essentially different from the use fo a coordinate frame, whether rectangular or, more generally, oblique. Distances measured along the diameter from the point of tangency are the abscissas, and segments parallel to the tangent and intercepted between the axis and the curve are the ordinates. The Apollonian relationship between these abscissas and the corresponding ordinates are nothing more nor less than rhetorical forms of the equations of the curves. However, Greek geometric algebra did not provide for negative magnitudes; moreover, the coordinate system was in every case superimposed *a posteriori* upon a given curve in order to study its properties. There appear to be no cases in ancient geometry in which a coordinate frame of reference was laid down *a priori* for purposes of graphical representation of an equation or relationship, whether symbolically or rhetorically expressed. Of Greek geometry we may say that equations are determined by curves, but not that curves are determined by equations. Coordinates, variables, and equations were subsidiary notions derived from a specific geometric situation; [...] That Apollonius, the greatest geometer of antiquity, failed to develop analytic geometry, was probably the result of a poverty of curves rather than of thought. General methods are not necessary when problems concern always one of a limited number of particular cases."

[2] Boyer, Carl B. (1991). "Apollonius of Perga". *A History of Mathematics* (Second ed.). John Wiley & Sons, Inc.. p. 142. ISBN 0-471-54397-7. "The Apollonian treatise *On Determinate Section* dealt with what might be called an analytic geometry of one dimension. It considered the following general problem, using the typical Greek algebraic analysis in geometric form: Given four points A, B, C, D on a straight line, determine a fifth point P on it such that the rectangle on AP and CP is in a given ratio to the rectangle on BP and DP. Here, too, the problem reduces easily to the solution of a quadratic; and, as in other cases, Apollonius treated the question exhaustively, including the limits of possibility and the number of solutions."

[3] Chisholm, Hugh, ed. (1911). "Apollonius of Perga". *Encyclopædia Britannica* (11th ed.). Cambridge University Press.

References

- Alhazen (1985). Hogendijk, J.P.. ed. *Ibn al-Haytham's Completion of the "Conics"*. New York: Springer Verlag.
- Apollonius. Apollonii *Pergaei quae Graece exstant cum commentariis antiquis*. Edited by I. L. Heiberg. 2 volumes. (Leipzig: Teubner, 1891/1893).
- Apollonius. *Apollonius of Perga Conics Books I–III*. Translated by R. Catesby Taliaferro. (Santa Fe: Green Lion Press, 1998).
- Apollonius. *Apollonius of Perga Conics Book IV*. Translated with introduction and notes by Michael N. Fried. (Santa Fe: Green Lion Press, 2002).
- Fried, Michael N.; Unguru, Sabetai (2001). *Apollonius of Perga's Conica: Text, Context, Subtext*. Leiden: Brill.
- Knorr, W. R. (1986). *Ancient Tradition of Geometric Problems*. Cambridge, MA: Birkhauser Boston.
- Neugebauer, Otto (1975). *A History of Ancient Mathematical Astronomy*. New York: Springer-Verlag.
- Pappus of Alexandria (1986). Jones, A. ed. *Book 7 of the "Collection"*. New York: Springer-Verlag.
- Toomer, G.J. (1970). "Apollonius of Perga". *Dictionary of Scientific Biography*. **1**. New York: Charles Scribner's Sons. pp. 179–193. ISBN 0-684-10114-9.
- Zeuthen, H. G. (1886). *Die Lehre von den Kegelschnitten im Altertum*. Copenhagen: Höst and Sohn.

The Works of Apollonius of Perga online

- Text in Classical Greek: PDF scans of Heiberg's edition of the Conic Sections of Apollonius of Perga, now in the public domain (http://www.wilbourhall.org/index.html#apollonius)
- In English translation: *Treatise on the Conic Sections* (http://www.wilbourhall.org/index.html#apollonius), trans. T.L. Heath

External links

- Conic Sections of Apollonius (http://mathdl.maa.org/convergence/1/?pa=content&sa=viewDocument&nodeId=196&bodyId=201) at Convergence (http://mathdl.maa.org/convergence/1/)
- O'Connor, John J.; Robertson, Edmund F., "Apollonius of Perga" (http://www-history.mcs.st-andrews.ac.uk/Biographies/Apollonius.html), *MacTutor History of Mathematics archive*, University of St Andrews.

This article incorporates text from a publication now in the public domain: T. L. Heath (1911). "Apollonius of Perga". In Chisholm, Hugh. *Encyclopædia Britannica* (11th ed.). Cambridge University Press.

Archimedes

Archimedes of Syracuse (Greek: Ἀρχιμήδης)	
 Archimedes Thoughtful by Fetti (1620)	
Born	c. 287 BCSyracuse, Sicily Magna Graecia
Died	c. 212 BC (aged around 75)Syracuse
Residence	Syracuse, Sicily
Fields	Mathematics physics engineering astronomy invention
Known for	Archimedes' principle Archimedes' screw hydrostatics levers infinitesimals

Archimedes of Syracuse (Greek: Ἀρχιμήδης; c. 287 BC – c. 212 BC) was a Greek mathematician, physicist, engineer, inventor, and astronomer.[1] Although few details of his life are known, he is regarded as one of the leading scientists in classical antiquity. Among his advances in physics are the foundations of hydrostatics, statics and an explanation of the principle of the lever. He is credited with designing innovative machines, including siege engines and the screw pump that bears his name. Modern experiments have tested claims that Archimedes designed machines capable of lifting attacking ships out of the water and setting ships on fire using an array of mirrors.[2]

Archimedes is generally considered to be the greatest mathematician of antiquity and one of the greatest of all time.[3] [4] He used the method of exhaustion to calculate the area under the arc of a parabola with the summation of an infinite series, and gave a remarkably accurate approximation of pi.[5] He also defined the spiral bearing his name, formulae for the volumes of surfaces of revolution and an ingenious system for expressing very large numbers.

Archimedes died during the Siege of Syracuse when he was killed by a Roman soldier despite orders that he should not be harmed. Cicero describes visiting the tomb of Archimedes, which was surmounted by a sphere inscribed within a cylinder. Archimedes had proven that the sphere has two thirds of the volume and surface area of the cylinder (including the bases of the latter), and regarded this as the greatest of his mathematical achievements.

Unlike his inventions, the mathematical writings of Archimedes were little known in antiquity. Mathematicians from Alexandria read and quoted him, but the first comprehensive compilation was not made until c. 530 AD by Isidore of Miletus, while commentaries on the works of Archimedes written by Eutocius in the sixth century AD opened them to wider readership for the first time. The relatively few copies of Archimedes' written work that survived through the Middle Ages were an influential source of ideas for scientists during the Renaissance,[6] while the discovery in 1906 of previously unknown works by Archimedes in the Archimedes Palimpsest has provided new insights into how he obtained mathematical results.[7]

Biography

This bronze statue of Archimedes is at the Archenhold Observatory in Berlin. It was sculpted by Gerhard Thieme and unveiled in 1972.

Archimedes was born *c.* 287 BC in the seaport city of Syracuse, Sicily, at that time a self-governing colony in Magna Graecia. The date of birth is based on a statement by the Byzantine Greek historian John Tzetzes that Archimedes lived for 75 years.[8] In *The Sand Reckoner*, Archimedes gives his father's name as Phidias, an astronomer about whom nothing is known. Plutarch wrote in his *Parallel Lives* that Archimedes was related to King Hiero II, the ruler of Syracuse.[9] A biography of Archimedes was written by his friend Heracleides but this work has been lost, leaving the details of his life obscure.[10] It is unknown, for instance, whether he ever married or had children. During his youth, Archimedes may have studied in Alexandria, Egypt, where Conon of Samos and Eratosthenes of Cyrene were contemporaries. He referred to Conon of Samos as his friend, while two of his works (*The Method of Mechanical Theorems* and the *Cattle Problem*) have introductions addressed to Eratosthenes.[a]

Archimedes died *c.* 212 BC during the Second Punic War, when Roman forces under General Marcus Claudius Marcellus captured the city of Syracuse after a two-year-long siege. According to the popular account given by Plutarch, Archimedes was contemplating a mathematical diagram when the city was captured. A Roman soldier commanded him to come and meet General Marcellus but he declined, saying that he had to finish working on the problem. The soldier was enraged by this, and killed Archimedes with his sword. Plutarch also gives a lesser-known account of the death of Archimedes which suggests that he may have been killed while attempting to surrender to a Roman soldier. According to this story, Archimedes was carrying mathematical instruments, and was killed because the soldier thought that they were valuable items. General Marcellus was reportedly angered by the death of Archimedes, as he considered him a valuable scientific asset and had ordered that he not be harmed.[11]

The last words attributed to Archimedes are "Do not disturb my circles" (Greek: μή μου τοὺς κύκλους τάραττε), a reference to the circles in the mathematical drawing that he was supposedly studying when disturbed by the Roman soldier. This quote is often given in Latin as "Noli turbare circulos meos," but there is no reliable evidence that Archimedes uttered these words and they do not appear in the account given by Plutarch.[11]

The tomb of Archimedes carried a sculpture illustrating his favorite mathematical proof, consisting of a sphere and a cylinder of the same height and diameter. Archimedes had proven that the volume and surface area of the sphere are two thirds that of the cylinder including its bases. In 75 BC, 137 years after his death, the Roman orator Cicero was serving as quaestor in Sicily. He had heard stories about the tomb of Archimedes, but none of the locals was able to give him the location. Eventually he found the tomb near the Agrigentine gate in Syracuse, in a neglected condition and overgrown with bushes. Cicero had the tomb cleaned up, and was able to see the carving and read some of the verses that had been added as an inscription.[12] A tomb discovered in a hotel courtyard in Syracuse in the early 1960s was claimed to be that of Archimedes, but its location today is unknown.[13]

The standard versions of the life of Archimedes were written long after his death by the historians of Ancient Rome. The account of the siege of Syracuse given by Polybius in his *Universal History* was written around seventy years after Archimedes' death, and was used subsequently as a source by Plutarch and Livy. It sheds little light on Archimedes as a person, and focuses on the war machines that he is said to have built in order to defend the city.[14]

Discoveries and inventions

Archimedes' principle

The most widely known anecdote about Archimedes tells of how he invented a method for determining the volume of an object with an irregular shape. According to Vitruvius, a votive crown for a temple had been made for King Hiero II, who had supplied the pure gold to be used, and Archimedes was asked to determine whether some silver had been substituted by the dishonest goldsmith.[15] Archimedes had to solve the problem without damaging the crown, so he could not melt it down into a regularly shaped body in order to calculate its density. While taking a bath, he noticed that the level of the water in the tub rose as he got in, and realized that this effect could be used to determine the volume of the crown. For practical purposes water is incompressible,[16] so the submerged crown would displace an amount of water equal to its own volume. By dividing the mass of the crown by the volume of water displaced, the density of the crown could be obtained. This density would be lower than that of gold if cheaper and less dense metals had been added. Archimedes then took to the streets naked, so excited by his discovery that he had forgotten to dress, crying "Eureka!" (Greek: "εὕρηκα!," meaning "I have found it!"). The test was conducted successfully, proving that silver had indeed been mixed in.[17]

The story of the golden crown does not appear in the known works of Archimedes. Moreover, the practicality of the method it describes has been called into question, due to the extreme accuracy with which one would have to measure the water displacement.[18] Archimedes may have instead sought a solution that applied the principle known in hydrostatics as Archimedes' principle, which he describes in his treatise *On Floating Bodies*. This principle states that a body immersed in a fluid experiences a buoyant force equal to the weight of the fluid it displaces.[19] Using this principle, it would have been possible to compare the density of the golden crown to that of solid gold by balancing the crown on a scale with a gold reference sample, then immersing the apparatus in water. The difference in density between the two samples would cause the scale to tip accordingly. Galileo considered it "probable that this method is the same that Archimedes followed, since, besides being very accurate, it is based on demonstrations found by Archimedes himself."[20]

Archimedes' screw

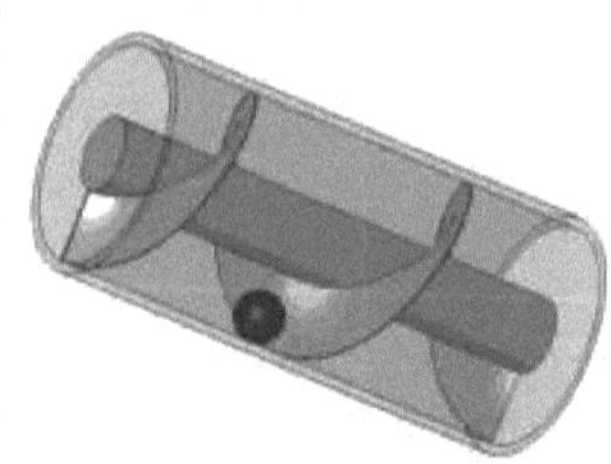

The Archimedes screw can raise water efficiently.

A large part of Archimedes' work in engineering arose from fulfilling the needs of his home city of Syracuse. The Greek writer Athenaeus of Naucratis described how King Hiero II commissioned Archimedes to design a huge ship, the *Syracusia*, which could be used for luxury travel, carrying supplies, and as a naval warship. The *Syracusia* is said to have been the largest ship built in classical antiquity.[21] According to Athenaeus, it was capable of carrying 600 people and included garden decorations, a gymnasium and a temple dedicated to the goddess Aphrodite among its facilities. Since a ship of this size would leak a considerable amount of water through the hull, the Archimedes screw was purportedly developed in order to remove the bilge water. Archimedes' machine was a device with a revolving screw-shaped blade inside a cylinder. It was turned by hand, and could also be used to transfer water from a low-lying body of water into irrigation canals. The Archimedes screw is still in use today for pumping liquids and granulated solids such as coal and grain. The Archimedes screw described in Roman times by Vitruvius may have been an improvement on a screw pump that was used to irrigate the Hanging Gardens of Babylon.[22] [23] [24] The world's first seagoing steamship with a screw propeller was the *SS Archimedes*, which was launched in 1839 and named in honor of Archimedes and his work on the screw.[25]

Claw of Archimedes

The Claw of Archimedes is a weapon that he is said to have designed in order to defend the city of Syracuse. Also known as "the ship shaker," the claw consisted of a crane-like arm from which a large metal grappling hook was suspended. When the claw was dropped onto an attacking ship the arm would swing upwards, lifting the ship out of the water and possibly sinking it. There have been modern experiments to test the feasibility of the claw, and in 2005 a television documentary entitled *Superweapons of the Ancient World* built a version of the claw and concluded that it was a workable device.[26] [27]

Heat ray

The 2nd century AD author Lucian wrote that during the Siege of Syracuse (*c.* 214–212 BC), Archimedes destroyed enemy ships with fire. Centuries later, Anthemius of Tralles mentions burning-glasses as Archimedes' weapon.[28] The device, sometimes called the "Archimedes heat ray", was used to focus sunlight onto approaching ships, causing them to catch fire.

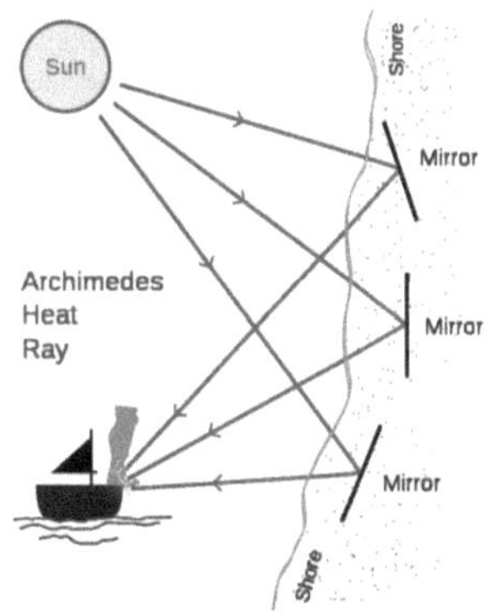

Archimedes may have used mirrors acting collectively as a parabolic reflector to burn ships attacking Syracuse.

This purported weapon has been the subject of ongoing debate about its credibility since the Renaissance. René Descartes rejected it as false, while modern researchers have attempted to recreate the effect using only the means that would have been available to Archimedes.[29] It has been suggested that a large array of highly polished bronze or copper shields acting as mirrors could have been employed to focus sunlight onto a ship. This would have used the principle of the parabolic reflector in a manner similar to a solar furnace.

A test of the Archimedes heat ray was carried out in 1973 by the Greek scientist Ioannis Sakkas. The experiment took place at the Skaramagas naval base outside Athens. On this occasion 70 mirrors were used, each with a copper coating and a size of around five by three feet (1.5 by 1 m). The mirrors were pointed at a plywood mock-up of a Roman warship at a distance of around 160 feet (50 m). When the mirrors were focused accurately, the ship burst into flames within a few seconds. The plywood ship had a coating of tar paint, which may have aided combustion.[30] A coating of tar would have been commonplace on ships in the classical era.[d]

In October 2005 a group of students from the Massachusetts Institute of Technology carried out an experiment with 127 one-foot (30 cm) square mirror tiles, focused on a mock-up wooden ship at a range of around 100 feet (30 m). Flames broke out on a patch of the ship, but only after the sky had been cloudless and the ship had remained stationary for around ten minutes. It was concluded that the device was a feasible weapon under these conditions. The MIT group repeated the experiment for the television show *MythBusters*, using a wooden fishing boat in San Francisco as the target. Again some charring occurred, along with a small amount of flame. In order to catch fire, wood needs to reach its autoignition temperature, which is around 300 °C (570 °F).[31] [32]

When *MythBusters* broadcast the result of the San Francisco experiment in January 2006, the claim was placed in the category of "busted" (or failed) because of the length of time and the ideal weather conditions required for combustion to occur. It was also pointed out that since Syracuse faces the sea towards the east, the Roman fleet would have had to attack during the morning for optimal gathering of light by the mirrors. *MythBusters* also pointed out that conventional weaponry, such as flaming arrows or bolts from a catapult, would have been a far easier way of setting a ship on fire at short distances.[2]

In December 2010, *MythBusters* again looked at the heat ray story in a special edition featuring Barack Obama, entitled *President's Challenge*. Several experiments were carried out, including a large scale test with 500 schoolchildren aiming mirrors at a mock-up of a Roman sailing ship 400 feet (120 m) away. In all of the experiments, the sail failed to reach the 210 °C (410 °F) required to catch fire, and the verdict was again "busted". The show concluded that a more likely effect of the mirrors would have been blinding, dazzling, or distracting the crew of the ship.[33]

Other discoveries and inventions

While Archimedes did not invent the lever, he gave an explanation of the principle involved in his work *On the Equilibrium of Planes*. Earlier descriptions of the lever are found in the Peripatetic school of the followers of Aristotle, and are sometimes attributed to Archytas.[34] [35] According to Pappus of Alexandria, Archimedes' work on levers caused him to remark: "Give me a place to stand on, and I will move the Earth." (Greek: δῶς μοι πᾶ στῶ καὶ τὰν γᾶν κινάσω)[36] Plutarch describes how Archimedes designed block-and-tackle pulley systems, allowing sailors to use the principle of leverage to lift objects that would otherwise have been too heavy to move.[37] Archimedes has also been credited with improving the power and accuracy of the catapult, and with inventing the odometer during the First Punic War. The odometer was described as a cart with a gear mechanism that dropped a ball into a container after each mile traveled.[38]

Cicero (106–43 BC) mentions Archimedes briefly in his dialogue *De re publica*, which portrays a fictional conversation taking place in 129 BC. After the capture of Syracuse *c.* 212 BC, General Marcus Claudius Marcellus is said to have taken back to Rome two mechanisms, constructed by Archimedes and used as aids in astronomy, which showed the motion of the Sun, Moon and five planets. Cicero mentions similar mechanisms designed by Thales of Miletus and Eudoxus of Cnidus. The dialogue says that Marcellus kept one of the devices as his only personal loot from Syracuse, and donated the other to the Temple of Virtue in Rome. Marcellus' mechanism was demonstrated, according to Cicero, by Gaius Sulpicius Gallus to Lucius Furius Philus, who described it thus:

> Hanc sphaeram Gallus cum moveret, fiebat ut soli luna totidem conversionibus in aere illo quot diebus in ipso caelo succederet, ex quo et in caelo sphaera solis fieret eadem illa defectio, et incideret luna tum in eam metam quae esset umbra terrae, cum sol e regione. — When Gallus moved the globe, it happened that the Moon followed the Sun by as many turns on that bronze contrivance as in the sky itself, from which also in the sky the Sun's globe became to have that same eclipse, and the Moon came then to that position which was its shadow on the Earth, when the Sun was in line.[39] [40]

This is a description of a planetarium or orrery. Pappus of Alexandria stated that Archimedes had written a manuscript (now lost) on the construction of these mechanisms entitled *On Sphere-Making*. Modern research in this area has been focused on the Antikythera mechanism, another device from classical antiquity that was probably designed for the same purpose. Constructing mechanisms of this kind would have required a sophisticated knowledge of differential gearing. This was once thought to have been beyond the range of the technology available in ancient times, but the discovery of the Antikythera mechanism in 1902 has confirmed that devices of this kind were known to the ancient Greeks.[41] [42]

Mathematics

While he is often regarded as a designer of mechanical devices, Archimedes also made contributions to the field of mathematics. Plutarch wrote: "He placed his whole affection and ambition in those purer speculations where there can be no reference to the vulgar needs of life."[43]

Archimedes was able to use infinitesimals in a way that is similar to modern integral calculus. Through proof by contradiction (reductio ad absurdum), he could give answers to problems to an arbitrary degree of accuracy, while specifying the limits within which the answer lay. This technique is known as the method of exhaustion, and he employed it to approximate the value of π. In *Measurement of a Circle* he did this by drawing a larger regular hexagon outside a circle and a smaller regular hexagon inside the circle, and progressively doubling the number of sides of each regular polygon, calculating the length of a side of each polygon at each step. As the number of sides increases, it becomes a more accurate approximation of a circle. After four such steps, when the polygons had 96 sides each, he was able to determine that the value of π lay between $3\frac{1}{7}$ (approximately 3.1429) and $3\frac{10}{71}$ (approximately 3.1408), consistent with its actual value of approximately 3.1416.[44] He also proved that the area of a circle was equal to π multiplied by the square of the radius of the circle (πr^2). In *On the Sphere and Cylinder*, Archimedes postulates that any magnitude when added to itself enough times will exceed any given magnitude. This is the Archimedean property of real numbers.[45]

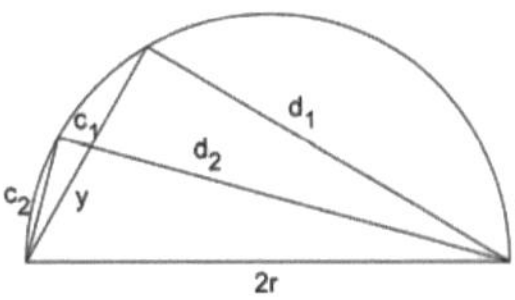

Archimedes used Pythagoras' Theorem to calculate the side of the 12-gon from that of the hexagon and for each subsequent doubling of the sides of the regular polygon.

In *Measurement of a Circle*, Archimedes gives the value of the square root of 3 as lying between $\frac{265}{153}$ (approximately 1.7320261) and $\frac{1351}{780}$ (approximately 1.7320512). The actual value is approximately 1.7320508, making this a very accurate estimate. He introduced this result without offering any explanation of how he had obtained it. This aspect of the work of Archimedes caused John Wallis to remark that he was: "as it were of set purpose to have covered up the traces of his investigation as if he had grudged posterity the secret of his method of inquiry while he wished to extort from them assent to his results."[46] It is possible that he used an iterative procedure to calculate these values.[47]

In *The Quadrature of the Parabola*, Archimedes proved that the area enclosed by a parabola and a straight line is $\frac{4}{3}$ times the area of a corresponding inscribed triangle as shown in the figure at right. He expressed the solution to the problem as an infinite geometric series with the common ratio $\frac{1}{4}$:

If the first term in this series is the area of the triangle, then the second is the sum of the areas of two triangles whose bases are the two smaller secant lines, and so on. This proof uses a variation of the series 1/4 + 1/16 + 1/64 + 1/256 + · · · which sums to $\frac{1}{3}$.

In *The Sand Reckoner*, Archimedes set out to calculate the number of grains of sand that the universe could contain. In doing so, he challenged the notion that the number of grains of sand was too large to be counted. He wrote: "There are some, King Gelo (Gelo II, son of Hiero II), who think that the number of the sand is infinite in multitude; and I mean by the sand not only that which exists about Syracuse and the rest of Sicily but also that which is found in every region whether inhabited or uninhabited." To solve the problem, Archimedes devised a system of counting based on the myriad. The word is from the Greek μυριάς *murias*, for the number 10,000. He proposed a number system using powers of a myriad of myriads (100 million) and concluded that the number of grains of sand required to fill the universe would be 8 vigintillion, or 8×10^{63}.[48]

Writings

The works of Archimedes were written in Doric Greek, the dialect of ancient Syracuse.[49] The written work of Archimedes has not survived as well as that of Euclid, and seven of his treatises are known to have existed only through references made to them by other authors. Pappus of Alexandria mentions *On Sphere-Making* and another work on polyhedra, while Theon of Alexandria quotes a remark about refraction from the now-lost *Catoptrica*.[b] During his lifetime, Archimedes made his work known through correspondence with the mathematicians in Alexandria. The writings of Archimedes were collected by the Byzantine architect Isidore of Miletus (*c.* 530 AD), while commentaries on the works of Archimedes written by Eutocius in the sixth century AD helped to bring his work a wider audience. Archimedes' work was translated into Arabic by Thābit ibn Qurra (836–901 AD), and Latin by Gerard of Cremona (*c.* 1114–1187 AD). During the Renaissance, the *Editio Princeps* (First Edition) was published in Basel in 1544 by Johann Herwagen with the works of Archimedes in Greek and Latin.[50] Around the year 1586 Galileo Galilei invented a hydrostatic balance for weighing metals in air and water after apparently being inspired by the work of Archimedes.[51]

Surviving works

- *On the Equilibrium of Planes* (two volumes)

  The first book is in fifteen propositions with seven postulates, while the second book is in ten propositions. In this work Archimedes explains the *Law of the Lever*, stating, "Magnitudes are in equilibrium at distances reciprocally proportional to their weights."

  Archimedes uses the principles derived to calculate the areas and centers of gravity of various geometric figures including triangles, parallelograms and parabolas.[52]

- *On the Measurement of a Circle*

  This is a short work consisting of three propositions. It is written in the form of a correspondence with Dositheus of Pelusium, who was a student of Conon of Samos. In Proposition II, Archimedes gives an approximation of the value of pi (π), showing that it is greater than $^{223}/_{71}$ and less than $^{22}/_{7}$.

- *On Spirals*

  This work of 28 propositions is also addressed to Dositheus. The treatise defines what is now called the Archimedean spiral. It is the locus of points corresponding to the locations over time of a point moving away from a fixed point with a constant speed along a line which rotates with constant angular velocity. Equivalently, in polar coordinates (r, θ) it can be described by the equation

  with real numbers a and b. This is an early example of a mechanical curve (a curve traced by a moving point) considered by a Greek mathematician.

- *On the Sphere and the Cylinder* (two volumes)

  In this treatise addressed to Dositheus, Archimedes obtains the result of which he was most proud, namely the relationship between a sphere and a circumscribed cylinder of the same height and diameter. The volume is $^4/_3\pi r^3$ for the sphere, and $2\pi r^3$ for the cylinder. The surface area is $4\pi r^2$ for the sphere, and $6\pi r^2$ for the cylinder (including its two bases), where r is the radius of the sphere and cylinder. The sphere has a volume two-thirds that of the circumscribed cylinder. Similarly, the sphere has an area two-thirds that of the cylinder (including the bases). A sculpted sphere and cylinder were placed on the tomb of Archimedes at his request.

- *On Conoids and Spheroids*

  This is a work in 32 propositions addressed to Dositheus. In this treatise Archimedes calculates the areas and volumes of sections of cones, spheres, and paraboloids.

- *On Floating Bodies* (two volumes)

In the first part of this treatise, Archimedes spells out the law of equilibrium of fluids, and proves that water will adopt a spherical form around a center of gravity. This may have been an attempt at explaining the theory of contemporary Greek astronomers such as Eratosthenes that the Earth is round. The fluids described by Archimedes are not self-gravitating, since he assumes the existence of a point towards which all things fall in order to derive the spherical shape.

In the second part, he calculates the equilibrium positions of sections of paraboloids. This was probably an idealization of the shapes of ships' hulls. Some of his sections float with the base under water and the summit above water, similar to the way that icebergs float. Archimedes' principle of buoyancy is given in the work, stated as follows:

> Any body wholly or partially immersed in a fluid experiences an upthrust equal to, but opposite in sense to, the weight of the fluid displaced.

- *The Quadrature of the Parabola*

  In this work of 24 propositions addressed to Dositheus, Archimedes proves by two methods that the area enclosed by a parabola and a straight line is 4/3 multiplied by the area of a triangle with equal base and height. He achieves this by calculating the value of a geometric series that sums to infinity with the ratio $^1/_4$.

- *(O)stomachion*

  This is a dissection puzzle similar to a Tangram, and the treatise describing it was found in more complete form in the Archimedes Palimpsest. Archimedes calculates the areas of the 14 pieces which can be assembled to form a square. Research published by Dr. Reviel Netz of Stanford University in 2003 argued that Archimedes was attempting to determine how many ways the pieces could be assembled into the shape of a square. Dr. Netz calculates that the pieces can be made into a square 17,152 ways.[53] The number of arrangements is 536 when solutions that are equivalent by rotation and reflection have been excluded.[54] The puzzle represents an example of an early problem in combinatorics.

  The origin of the puzzle's name is unclear, and it has been suggested that it is taken from the Ancient Greek word for throat or gullet, stomachos (στόμαχος).[55] Ausonius refers to the puzzle as *Ostomachion*, a Greek compound word formed from the roots of ὀστέον (*osteon*, bone) and μάχη (machē – fight). The puzzle is also known as the Loculus of Archimedes or Archimedes' Box.[56]

- *Archimedes' cattle problem*

  This work was discovered by Gotthold Ephraim Lessing in a Greek manuscript consisting of a poem of 44 lines, in the Herzog August Library in Wolfenbüttel, Germany in 1773. It is addressed to Eratosthenes and the mathematicians in Alexandria. Archimedes challenges them to count the numbers of cattle in the Herd of the Sun by solving a number of simultaneous Diophantine equations. There is a more difficult version of the problem in which some of the answers are required to be square numbers. This version of the problem was first solved by A. Amthor[57] in 1880, and the answer is a very large number, approximately $7.760271\times10^{206544}$.[58]

- *The Sand Reckoner*

  In this treatise, Archimedes counts the number of grains of sand that will fit inside the universe. This book mentions the heliocentric theory of the solar system proposed by Aristarchus of Samos, as well as contemporary ideas about the size of the Earth and the distance between various celestial bodies. By using a system of numbers based on powers of the myriad, Archimedes concludes that the number of grains of sand required to fill the universe is 8×10^{63} in modern notation. The introductory letter states that Archimedes' father was an astronomer named Phidias. *The Sand Reckoner* or *Psammites* is the only surviving work in which Archimedes discusses his views on astronomy.[59]

- *The Method of Mechanical Theorems*

This treatise was thought lost until the discovery of the Archimedes Palimpsest in 1906. In this work Archimedes uses infinitesimals, and shows how breaking up a figure into an infinite number of infinitely small parts can be used to determine its area or volume. Archimedes may have considered this method lacking in formal rigor, so he also used the method of exhaustion to derive the results. As with *The Cattle Problem*, *The Method of Mechanical Theorems* was written in the form of a letter to Eratosthenes in Alexandria.

Apocryphal works

Archimedes' *Book of Lemmas* or *Liber Assumptorum* is a treatise with fifteen propositions on the nature of circles. The earliest known copy of the text is in Arabic. The scholars T. L. Heath and Marshall Clagett argued that it cannot have been written by Archimedes in its current form, since it quotes Archimedes, suggesting modification by another author. The *Lemmas* may be based on an earlier work by Archimedes that is now lost.[60]

It has also been claimed that Heron's formula for calculating the area of a triangle from the length of its sides was known to Archimedes.[c] However, the first reliable reference to the formula is given by Heron of Alexandria in the 1st century AD.[61]

Archimedes Palimpsest

The foremost document containing the work of Archimedes is the Archimedes Palimpsest. In 1906, the Danish professor Johan Ludvig Heiberg visited Constantinople and examined a 174-page goatskin parchment of prayers written in the 13th century AD. He discovered that it was a palimpsest, a document with text that had been written over an erased older work. Palimpsests were created by scraping the ink from existing works and reusing them, which was a common practice in the Middle Ages as vellum was expensive. The older works in the palimpsest were identified by scholars as 10th century AD copies of previously unknown treatises by Archimedes.[62] The parchment spent hundreds of years in a monastery library in Constantinople before being sold to a private collector in the 1920s. On October 29, 1998 it was sold at auction to an anonymous buyer for $2 million at Christie's in New York.[63] The palimpsest holds seven treatises, including the only surviving copy of *On Floating Bodies* in the original Greek. It is the only known source of *The Method of Mechanical Theorems*, referred to by Suidas and thought to have been lost forever. *Stomachion* was also discovered in the palimpsest, with a more complete analysis of the puzzle than had been found in previous texts. The palimpsest is now stored at the Walters Art Museum in Baltimore, Maryland, where it has been subjected to a range of modern tests including the use of ultraviolet and x-ray light to read the overwritten text.[64]

Stomachion is a dissection puzzle in the Archimedes Palimpsest.

The treatises in the Archimedes Palimpsest are: *On the Equilibrium of Planes, On Spirals, Measurement of a Circle, On the Sphere and the Cylinder, On Floating Bodies, The Method of Mechanical Theorems* and *Stomachion.*

Legacy

- There is a crater on the Moon named Archimedes (29.7° N, 4.0° W) in his honor, as well as a lunar mountain range, the Montes Archimedes (25.3° N, 4.6° W).[65]
- The asteroid 3600 Archimedes is named after him.[66]
- The Fields Medal for outstanding achievement in mathematics carries a portrait of Archimedes, along with a carving illustrating his proof on the sphere and the cylinder. The inscription around the head of Archimedes is a quote attributed to him which reads in Latin: "Transire suum pectus mundoque potiri" (Rise above oneself and grasp the world).[67]
- Archimedes has appeared on postage stamps issued by East Germany (1973), Greece (1983), Italy (1983), Nicaragua (1971), San Marino (1982), and Spain (1963).[68]
- The exclamation of Eureka! attributed to Archimedes is the state motto of California. In this instance the word refers to the discovery of gold near Sutter's Mill in 1848 which sparked the California Gold Rush.[69]
- A movement for civic engagement targeting universal access to health care in the US state of Oregon has been named the "Archimedes Movement," headed by former Oregon Governor John Kitzhaber.[70]

See also

• Arbelos	• Archimedes' screw	• Diocles	• Salinon
• Archimedes' axiom	• Archimedean solid	• List of things named after Archimedes	• Steam cannon
• Archimedes number	• Archimedes' twin circles	• Methods of computing square roots	• Syracusia
• Archimedes paradox	• Archimedes' use of infinitesimals	• Pseudo-Archimedes	• Vitruvius
• Archimedes principle of buoyancy	• Archytas		• Zhang Heng

Notes and references

Notes

a. In the preface to *On Spirals* addressed to Dositheus of Pelusium, Archimedes says that "many years have elapsed since Conon's death." Conon of Samos lived *c.* 280–220 BC, suggesting that Archimedes may have been an older man when writing some of his works.

b. The treatises by Archimedes known to exist only through references in the works of other authors are: *On Sphere-Making* and a work on polyhedra mentioned by Pappus of Alexandria; *Catoptrica*, a work on optics mentioned by Theon of Alexandria; *Principles*, addressed to Zeuxippus and explaining the number system used in *The Sand Reckoner*; *On Balances and Levers*; *On Centers of Gravity*; *On the Calendar*. Of the surviving works by Archimedes, T. L. Heath offers the following suggestion as to the order in which they were written: *On the Equilibrium of Planes I*, *The Quadrature of the Parabola*, *On the Equilibrium of Planes II*, *On the Sphere and the Cylinder I, II*, *On Spirals*, *On Conoids and Spheroids*, *On Floating Bodies I, II*, *On the Measurement of a Circle*, *The Sand Reckoner*.

c. Boyer, Carl Benjamin *A History of Mathematics* (1991) ISBN 0-471-54397-7 "Arabic scholars inform us that the familiar area formula for a triangle in terms of its three sides, usually known as Heron's formula — $k = \sqrt{(s(s-a)(s-b)(s-c))}$, where s is the semiperimeter — was known to Archimedes several centuries before Heron lived. Arabic scholars also attribute to Archimedes the 'theorem on the broken chord' ... Archimedes is reported by the Arabs to have given several proofs of the theorem."

d. "It was usual to smear the seams or even the whole hull with pitch or with pitch and wax". In Νεκρικοὶ Διάλογοι (*Dialogues of the Dead*), Lucian refers to coating the seams of a skiff with wax, a reference to pitch (tar) or wax.[71]

References

[1] "Archimedes (c.287 - c.212 BC)" (http://www.bbc.co.uk/history/historic_figures/archimedes.shtml). *BBC History*. . Retrieved 2012-06-07.

[2] "Archimedes Death Ray: Testing with MythBusters" (http://web.mit.edu/2.009/www//experiments/deathray/10_Mythbusters.html). MIT. . Retrieved 2007-07-23.

[3] Calinger, Ronald (1999). *A Contextual History of Mathematics*. Prentice-Hall. p. 150. ISBN 0-02-318285-7. "Shortly after Euclid, compiler of the definitive textbook, came Archimedes of Syracuse (ca. 287 212 BC), the most original and profound mathematician of antiquity."

[4] "Archimedes of Syracuse" (http://www-history.mcs.st-and.ac.uk/Biographies/Archimedes.html). The MacTutor History of Mathematics archive. January 1999. . Retrieved 2008-06-09.

[5] O'Connor, J.J. and Robertson, E.F. (February 1996). "A history of calculus" (http://www-groups.dcs.st-and.ac.uk/~history/HistTopics/The_rise_of_calculus.html). University of St Andrews. Archived (http://web.archive.org/web/20070715191704/http://www-groups.dcs.st-and.ac.uk/~history/HistTopics/The_rise_of_calculus.html) from the original on 15 July 2007. . Retrieved 2007-08-07.

[6] Bursill-Hall, Piers. "Galileo, Archimedes, and Renaissance engineers" (http://www.sciencelive.org/component/option,com_mediadb/task,view/idstr,CU-MMP-PiersBursillHall/Itemid,30). sciencelive with the University of Cambridge. . Retrieved 2007-08-07.

[7] "Archimedes – The Palimpsest" (http://web.archive.org/web/20070928102802/http://www.archimedespalimpsest.org/palimpsest_making1.html). Walters Art Museum. Archived from the original (http://www.archimedespalimpsest.org/palimpsest_making1.html) on 2007-09-28. . Retrieved 2007-10-14.

[8] Heath, T. L., *Works of Archimedes*, 1897

[9] Plutarch. "*Parallel Lives* Complete e-text from Gutenberg.org" (http://www.gutenberg.org/etext/674). Project Gutenberg. Archived (http://web.archive.org/web/20070711045124/http://www.gutenberg.org/etext/674) from the original on 11 July 2007. . Retrieved 2007-07-23.

[10] O'Connor, J.J. and Robertson, E.F.. "Archimedes of Syracuse" (http://www-history.mcs.st-andrews.ac.uk/Biographies/Archimedes.html). University of St Andrews. Archived (http://web.archive.org/web/20070206082010/http://www-history.mcs.st-andrews.ac.uk/Biographies/Archimedes.html) from the original on 6 February 2007. . Retrieved 2007-01-02.

[11] Rorres, Chris. "Death of Archimedes: Sources" (http://www.math.nyu.edu/~crorres/Archimedes/Death/Histories.html). Courant Institute of Mathematical Sciences. Archived (http://web.archive.org/web/20061210060235/http://www.math.nyu.edu/~crorres/Archimedes/Death/Histories.html) from the original on 10 December 2006. . Retrieved 2007-01-02.

[12] Rorres, Chris. "Tomb of Archimedes: Sources" (http://www.math.nyu.edu/~crorres/Archimedes/Tomb/Cicero.html). Courant Institute of Mathematical Sciences. Archived (http://web.archive.org/web/20061209201723/http://www.math.nyu.edu/~crorres/Archimedes/Tomb/Cicero.html) from the original on 9 December 2006. . Retrieved 2007-01-02.

[13] Rorres, Chris. "Tomb of Archimedes – Illustrations" (http://www.math.nyu.edu/~crorres/Archimedes/Tomb/TombIllus.html). Courant Institute of Mathematical Sciences. . Retrieved 2011-03-15.

[14] Rorres, Chris. "Siege of Syracuse" (http://www.math.nyu.edu/~crorres/Archimedes/Siege/Polybius.html). Courant Institute of Mathematical Sciences. Archived (http://web.archive.org/web/20070609013114/http://www.math.nyu.edu/~crorres/Archimedes/Siege/Polybius.html) from the original on 9 June 2007. . Retrieved 2007-07-23.

[15] Vitruvius. "*De Architectura*, Book IX, paragraphs 9–12, text in English and Latin" (http://penelope.uchicago.edu/Thayer/E/Roman/Texts/Vitruvius/9*.html). University of Chicago. . Retrieved 2007-08-30.

[16] "Incompressibility of Water" (http://www.fas.harvard.edu/~scdiroff/lds/NewtonianMechanics/IncompressibilityofWater/IncompressibilityofWater.html). Harvard University. Archived (http://web.archive.org/web/20080317130651/http://www.fas.harvard.edu/~scdiroff/lds/NewtonianMechanics/IncompressibilityofWater/IncompressibilityofWater.html) from the original on 17 March 2008. . Retrieved 2008-02-27.

[17] HyperPhysics. "Buoyancy" (http://hyperphysics.phy-astr.gsu.edu/Hbase/pbuoy.html). Georgia State University. Archived (http://web.archive.org/web/20070714113647/http://hyperphysics.phy-astr.gsu.edu/hbase/pbuoy.html#c1) from the original on 14 July 2007. . Retrieved 2007-07-23.

[18] Rorres, Chris. "The Golden Crown" (http://www.math.nyu.edu/~crorres/Archimedes/Crown/CrownIntro.html). Drexel University. Archived (http://web.archive.org/web/20090311051318/http://www.math.nyu.edu/~crorres/Archimedes/Crown/CrownIntro.html) from the original on 11 March 2009. . Retrieved 2009-03-24.

[19] Carroll, Bradley W. "*Archimedes' Principle*" (http://www.physics.weber.edu/carroll/Archimedes/principle.htm). Weber State University. Archived (http://web.archive.org/web/20070808132323/http://physics.weber.edu/carroll/Archimedes/principle.htm) from the original on 8 August 2007. . Retrieved 2007-07-23.

[20] Rorres, Chris. "The Golden Crown: Galileo's Balance" (http://www.math.nyu.edu/~crorres/Archimedes/Crown/bilancetta.html). Drexel University. Archived (http://web.archive.org/web/20090224221137/http://math.nyu.edu/~crorres/Archimedes/Crown/bilancetta.html) from the original on 24 February 2009. . Retrieved 2009-03-24.

[21] Casson, Lionel (1971). *Ships and Seamanship in the Ancient World*. Princeton University Press. ISBN 0-691-03536-9.

[22] Dalley, Stephanie. Oleson, John Peter. "*Sennacherib, Archimedes, and the Water Screw: The Context of Invention in the Ancient World*" (http://muse.jhu.edu/journals/technology_and_culture/toc/tech44.1.html). *Technology and Culture* Volume 44, Number 1, January 2003 (PDF). . Retrieved 2007-07-23.

[23] Rorres, Chris. "Archimedes screw – Optimal Design" (http://www.cs.drexel.edu/~crorres/Archimedes/Screw/optimal/optimal.html). Courant Institute of Mathematical Sciences. . Retrieved 2007-07-23.

[24] An animation of an Archimedes screw

[25] "SS Archimedes" (http://www.wrecksite.eu/wreck.aspx?636). wrecksite.eu. . Retrieved 2011-01-22.

[26] Rorres, Chris. "Archimedes' Claw – Illustrations and Animations – a range of possible designs for the claw" (http://www.math.nyu.edu/~crorres/Archimedes/Claw/illustrations.html). Courant Institute of Mathematical Sciences. . Retrieved 2007-07-23.

[27] Carroll, Bradley W. "Archimedes' Claw – watch an animation" (http://physics.weber.edu/carroll/Archimedes/claw.htm). Weber State University. Archived (http://web.archive.org/web/20070813202716/http://physics.weber.edu/carroll/Archimedes/claw.htm) from the original on 13 August 2007. . Retrieved 2007-08-12.

[28] *Hippias*, 2 (cf. Galen, *On temperaments* 3.2, who mentions *pyreia*, "torches"); Anthemius of Tralles, *On miraculous engines* 153 [Westerman].

[29] John Wesley. "*A Compendium of Natural Philosophy* (1810) Chapter XII, *Burning Glasses*" (http://web.archive.org/web/20071012154432/http://wesley.nnu.edu/john_wesley/wesley_natural_philosophy/duten12.htm). Online text at Wesley Center for Applied Theology. Archived from the original (http://wesley.nnu.edu/john_wesley/wesley_natural_philosophy/duten12.htm) on 2007-10-12. . Retrieved 2007-09-14.

[30] "Archimedes' Weapon" (http://www.time.com/time/magazine/article/0,9171,908175,00.html?promoid=googlep). Time Magazine. November 26, 1973. . Retrieved 2007-08-12.

[31] Bonsor, Kevin. "How Wildfires Work" (http://science.howstuffworks.com/wildfire.htm). HowStuffWorks. Archived (http://web.archive.org/web/20070714174036/http://science.howstuffworks.com/wildfire.htm) from the original on 14 July 2007. . Retrieved 2007-07-23.

[32] Fuels and Chemicals – Auto Ignition Temperatures (http://www.engineeringtoolbox.com/fuels-ignition-temperatures-d_171.html)

[33] "TV Review: MythBusters 8.27 – President's Challenge" (http://fandomania.com/tv-review-mythbusters-8-27-presidents-challenge/). . Retrieved 2010-12-18.

[34] Rorres, Chris. "The Law of the Lever According to Archimedes" (http://www.math.nyu.edu/~crorres/Archimedes/Lever/LeverLaw.html). Courant Institute of Mathematical Sciences. . Retrieved 2010-03-20.

[35] Clagett, Marshall (2001). *Greek Science in Antiquity* (http://books.google.com/?id=mweWMAlf-tEC&pg=PA72&lpg=PA72&dq=archytas+lever&q=archytas lever). Dover Publications. ISBN 978-0-486-41973-2. . Retrieved 2010-03-20.

[36] Quoted by Pappus of Alexandria in *Synagoge*, Book VIII

[37] Dougherty, F. C.; Macari, J.; Okamoto, C.. "Pulleys" (http://www.swe.org/iac/lp/pulley_03.html). Society of Women Engineers. Archived (http://web.archive.org/web/20070718031943/http://www.swe.org/iac/LP/pulley_03.html) from the original on 18 July 2007. . Retrieved 2007-07-23.

[38] "Ancient Greek Scientists: Hero of Alexandria" (http://www.tmth.edu.gr/en/aet/5/55.html). Technology Museum of Thessaloniki. Archived (http://web.archive.org/web/20070905125400/http://www.tmth.edu.gr/en/aet/5/55.html) from the original on 5 September 2007. . Retrieved 2007-09-14.

[39] Cicero. "*De re publica* 1.xiv §21" (http://www.thelatinlibrary.com/cicero/repub1.shtml#21). thelatinlibrary.com. . Retrieved 2007-07-23.

[40] Cicero. "*De re publica* Complete e-text in English from Gutenberg.org" (http://www.gutenberg.org/etext/14988). Project Gutenberg. Archived (http://web.archive.org/web/20070929122153/http://www.gutenberg.org/etext/14988) from the original on 29 September 2007. . Retrieved 2007-09-18.

[41] Rorres, Chris. "Spheres and Planetaria" (http://www.math.nyu.edu/~crorres/Archimedes/Sphere/SphereIntro.html). Courant Institute of Mathematical Sciences. . Retrieved 2007-07-23.

[42] "Ancient Moon 'computer' revisited" (http://news.bbc.co.uk/1/hi/sci/tech/6191462.stm). BBC News. November 29, 2006. . Retrieved 2007-07-23.

[43] Plutarch. "Extract from *Parallel Lives*" (http://fulltextarchive.com/pages/Plutarch-s-Lives10.php#p35). fulltextarchive.com. . Retrieved 2009-08-10.

[44] Heath, T.L.. "Archimedes on measuring the circle" (http://www.math.ubc.ca/~cass/archimedes/circle.html). math.ubc.ca. . Retrieved 2012-10-30.

[45] Kaye, R.W.. "Archimedean ordered fields" (http://web.mat.bham.ac.uk/R.W.Kaye/seqser/archfields). web.mat.bham.ac.uk. . Retrieved 2009-11-07.

[46] Quoted in Heath, T. L. *Works of Archimedes*, Dover Publications, ISBN 0-486-42084-1.

[47] McKeeman, Bill. "The Computation of Pi by Archimedes" (http://www.mathworks.com/matlabcentral/fileexchange/29504-the-computation-of-pi-by-archimedes/content/html/ComputationOfPiByArchimedes.html#37). Matlab Central. . Retrieved 2012-10-30.

[48] Carroll, Bradley W. "The Sand Reckoner" (http://physics.weber.edu/carroll/Archimedes/sand.htm). Weber State University. Archived (http://web.archive.org/web/20070813215029/http://physics.weber.edu/carroll/Archimedes/sand.htm) from the original on 13 August 2007. . Retrieved 2007-07-23.

[49] Encyclopedia of ancient Greece By Wilson, Nigel Guy p. 77 (http://books.google.com/books?id=-aFtPdh6-2QC&pg=PA77) ISBN 0-7945-0225-3 (2006)

[50] "Editions of Archimedes' Work" (http://www.brown.edu/Facilities/University_Library/exhibits/math/wholefr.html). Brown University Library. Archived (http://web.archive.org/web/20070808235638/http://www.brown.edu/Facilities/University_Library/exhibits/math/wholefr.html) from the original on 8 August 2007. . Retrieved 2007-07-23.

[51] Van Helden, Al. "The Galileo Project: Hydrostatic Balance" (http://galileo.rice.edu/sci/instruments/balance.html). Rice University. Archived (http://web.archive.org/web/20070905185039/http://galileo.rice.edu/sci/instruments/balance.html) from the original on 5 September 2007. . Retrieved 2007-09-14.

[52] Heath, T.L.. "The Works of Archimedes (1897). The unabridged work in PDF form (19 MB)" (http://www.archive.org/details/worksofarchimede029517mbp). Archive.org. Archived (http://web.archive.org/web/20071006033058/http://www.archive.org/details/worksofarchimede029517mbp) from the original on 6 October 2007. . Retrieved 2007-10-14.

[53] Kolata, Gina (December 14, 2003). "In Archimedes' Puzzle, a New Eureka Moment" (http://query.nytimes.com/gst/fullpage.html?res=9D00E6DD133CF937A25751C1A9659C8B63&sec=&spon=&pagewanted=all). The New York Times. . Retrieved 2007-07-23.

[54] Ed Pegg Jr. (November 17, 2003). "The Loculus of Archimedes, Solved" (http://www.maa.org/editorial/mathgames/mathgames_11_17_03.html). Mathematical Association of America. Archived (http://web.archive.org/web/20080519094951/http://www.maa.org/editorial/mathgames/mathgames_11_17_03.html) from the original on 19 May 2008. . Retrieved 2008-05-18.

[55] Rorres, Chris. "Archimedes' Stomachion" (http://math.nyu.edu/~crorres/Archimedes/Stomachion/intro.html). Courant Institute of Mathematical Sciences. Archived (http://web.archive.org/web/20071026005336/http://www.math.nyu.edu/~crorres/Archimedes/Stomachion/intro.html) from the original on 26 October 2007. . Retrieved 2007-09-14.

[56] "Graeco Roman Puzzles" (http://www.archimedes-lab.org/latin.html#archimede). Gianni A. Sarcone and Marie J. Waeber. Archived (http://web.archive.org/web/20080514130547/http://www.archimedes-lab.org/latin.html) from the original on 14 May 2008. . Retrieved 2008-05-09.

[57] Krumbiegel, B. and Amthor, A. *Das Problema Bovinum des Archimedes*, Historisch-literarische Abteilung der Zeitschrift Für Mathematik und Physik 25 (1880) pp. 121–136, 153–171.

[58] Calkins, Keith G. "Archimedes' Problema Bovinum" (http://web.archive.org/web/20071012171254/http://andrews.edu/~calkins/profess/cattle.htm). Andrews University. Archived from the original (http://www.andrews.edu/~calkins/profess/cattle.htm) on 2007-10-12. . Retrieved 2007-09-14.

[59] "English translation of *The Sand Reckoner*" (http://www.math.uwaterloo.ca/navigation/ideas/reckoner.shtml). University of Waterloo. Archived (http://web.archive.org/web/20070811235335/http://www.math.uwaterloo.ca/navigation/ideas/reckoner.shtml) from the original on 11 August 2007. . Retrieved 2007-07-23.

[60] "Archimedes' Book of Lemmas" (http://www.cut-the-knot.org/Curriculum/Geometry/BookOfLemmas/index.shtml). cut-the-knot. Archived (http://web.archive.org/web/20070711111858/http://www.cut-the-knot.org/Curriculum/Geometry/BookOfLemmas/index.shtml) from the original on 11 July 2007. . Retrieved 2007-08-07.

[61] O'Connor, J.J. and Robertson, E.F. (April 1999). "Heron of Alexandria" (http://www-history.mcs.st-and.ac.uk/Biographies/Heron.html). University of St Andrews. . Retrieved 2010-02-17.

[62] Miller, Mary K. (March 2007). "Reading Between the Lines" (http://www.smithsonianmag.com/science-nature/archimedes.html). Smithsonian Magazine. Archived (http://web.archive.org/web/20080119024939/http://www.smithsonianmag.com/science-nature/archimedes.html?) from the original on 19 January 2008. . Retrieved 2008-01-24.

[63] "Rare work by Archimedes sells for $2 million" (http://web.archive.org/web/20080516000109/http://edition.cnn.com/books/news/9810/29/archimedes/). CNN. October 29, 1998. Archived from the original (http://edition.cnn.com/books/news/9810/29/archimedes/) on May 16, 2008. . Retrieved 2008-01-15.

[64] "X-rays reveal Archimedes' secrets" (http://news.bbc.co.uk/1/hi/sci/tech/5235894.stm). BBC News. August 2, 2006. Archived (http://web.archive.org/web/20070825091847/http://news.bbc.co.uk/1/hi/sci/tech/5235894.stm) from the original on 25 August 2007. . Retrieved 2007-07-23.

[65] Friedlander, Jay and Williams, Dave. "Oblique view of Archimedes crater on the Moon" (http://nssdc.gsfc.nasa.gov/imgcat/html/object_page/a15_m_1541.html). NASA. Archived (http://web.archive.org/web/20070819054033/http://nssdc.gsfc.nasa.gov/imgcat/html/object_page/a15_m_1541.html) from the original on 19 August 2007. . Retrieved 2007-09-13.

[66] "Planetary Data System" (http://starbrite.jpl.nasa.gov/pds-explorer/index.jsp?selection=othertarget&targname=3600 ARCHIMEDES). NASA. Archived (http://web.archive.org/web/20071012171730/http://starbrite.jpl.nasa.gov/pds-explorer/index.jsp?selection=othertarget&targname=3600+ARCHIMEDES) from the original on 12 October 2007. . Retrieved 2007-09-13.

[67] "Fields Medal" (http://web.archive.org/web/20070701033751/http://www.mathunion.org/medals/Fields/AboutPhotos.html). International Mathematical Union. Archived from the original (http://www.mathunion.org/medals/Fields/AboutPhotos.html) on July 1, 2007. . Retrieved 2007-07-23.

[68] Rorres, Chris. "Stamps of Archimedes" (http://math.nyu.edu/~crorres/Archimedes/Stamps/stamps.html). Courant Institute of Mathematical Sciences. . Retrieved 2007-08-25.

[69] "California Symbols" (http://www.capitolmuseum.ca.gov/VirtualTour.aspx?content1=1278&Content2=1374&Content3=1294). California State Capitol Museum. Archived (http://web.archive.org/web/20071012123245/http://capitolmuseum.ca.gov/VirtualTour.aspx?content1=1278&Content2=1374&Content3=1294) from the original on 12 October 2007. . Retrieved 2007-09-14.

[70] "The Archimedes Movement" (http://www.archimedesmovement.org/). .

[71] Casson, Lionel (1995). *Ships and seamanship in the ancient world* (http://books.google.com/books?id=sDpMh0gK2OUC&pg=PA18&dq=why+were+homer's+ships+black#v=onepage&q=why were homer's ships black&f=false). Baltimore: The Johns Hopkins University Press. pp. 211–212. ISBN 978-0-8018-5130-8. .

Further reading

- Boyer, Carl Benjamin (1991). *A History of Mathematics*. New York: Wiley. ISBN 0-471-54397-7.
- Clagett, Marshall (1964-1984). *Archimedes in the Middle Ages*. **5 vols**. Madison, WI: University of Wisconsin Press.
- Dijksterhuis, E.J. (1987). *Archimedes*. Princeton University Press, Princeton. ISBN 0-691-08421-1. Republished translation of the 1938 study of Archimedes and his works by an historian of science.
- Gow, Mary (2005). *Archimedes: Mathematical Genius of the Ancient World*. Enslow Publishers, Inc. ISBN 0-7660-2502-0.
- Hasan, Heather (2005). *Archimedes: The Father of Mathematics*. Rosen Central. ISBN 978-1-4042-0774-5.
- Heath, T.L. (1897). *Works of Archimedes*. Dover Publications. ISBN 0-486-42084-1. Complete works of Archimedes in English.
- Netz, Reviel and Noel, William (2007). *The Archimedes Codex*. Orion Publishing Group. ISBN 0-297-64547-1.
- Pickover, Clifford A. (2008). *Archimedes to Hawking: Laws of Science and the Great Minds Behind Them*. Oxford University Press. ISBN 978-0-19-533611-5.
- Simms, Dennis L. (1995). *Archimedes the Engineer*. Continuum International Publishing Group Ltd. ISBN 0-7201-2284-8.
- Stein, Sherman (1999). *Archimedes: What Did He Do Besides Cry Eureka?*. Mathematical Association of America. ISBN 0-88385-718-9.

The Works of Archimedes online

- Text in Classical Greek: PDF scans of Heiberg's edition of the Works of Archimedes, now in the public domain (http://www.wilbourhall.org)
- In English translation: *The Works of Archimedes* (http://www.archive.org/details/worksofarchimede029517mbp), trans. T.L. Heath; supplemented by *The Method of Mechanical Theorems* (http://books.google.com/books?id=suYGAAAAYAAJ), trans. L.G. Robinson

External links

- Archimedes (http://www.bbc.co.uk/programmes/b00773bv) on *In Our Time* at the BBC. (listen now (http://www.bbc.co.uk/iplayer/console/b00773bv/In_Our_Time_Archimedes))
- Archimedes (https://inpho.cogs.indiana.edu/thinker/2546) at the Indiana Philosophy Ontology Project
- Archimedes (http://philpapers.org/s/archimedes) at PhilPapers
- The Archimedes Palimpsest project at The Walters Art Museum in Baltimore, Maryland (http://www.archimedespalimpsest.org/)
- The Mathematical Achievements and Methodologies of Archimedes (http://mathdb.org/articles/archimedes/e_archimedes.htm)
- "Archimedes and the Square Root of 3" (http://www.mathpages.com/home/kmath038/kmath038.htm) at MathPages.com.
- "Archimedes on Spheres and Cylinders" (http://www.mathpages.com/home/kmath343/kmath343.htm) at MathPages.com.
- Photograph of the Sakkas experiment in 1973 (http://www.cs.drexel.edu/~crorres/bbc_archive/mirrors_sailors_sakas.jpg)
- Testing the Archimedes steam cannon (http://web.mit.edu/2.009/www/experiments/steamCannon/ArchimedesSteamCannon.html)
- Stamps of Archimedes (http://www.stampsbook.org/subject/Archimedes.html)
- Eureka! 1,000-year-old text by Greek maths genius Archimedes goes on display (http://www.dailymail.co.uk/sciencetech/article-2050631/Eureka-1-000-year-old-text-Greek-maths-genius-Archimedes-goes-display.html)

Daily Mail, October 18, 2011.

Diophantus

Diophantus of Alexandria (Ancient Greek: **Διόφαντος ὁ Ἀλεξανδρεύς**. b. between A.D. 200 and 214, d. between 284 and 298 at age 84), sometimes called "the father of algebra", was an Alexandrian Greek mathematician[1] [2] [3] [4] and the author of a series of books called *Arithmetica*. These texts deal with solving algebraic equations, many of which are now lost. In studying *Arithmetica,* Pierre de Fermat concluded that a certain equation considered by Diophantus had no solutions, and noted without elaboration that he had found "a truly marvelous proof of this proposition," now referred to as Fermat's Last Theorem. This led to tremendous advances in number theory, and the study of Diophantine equations ("Diophantine geometry") and of Diophantine approximations remain important areas of mathematical research. Diophantus coined the term παρισότης to refer to an approximate equality.[5] This term was rendered as *adaequalitat* in Latin, and became the technique of adequality developed by Pierre de Fermat to find maxima for functions and tangent lines to curves. Diophantus was the first Greek mathematician who recognized fractions as numbers; thus he allowed positive rational numbers for the coefficients and solutions. In modern use, Diophantine equations are usually algebraic equations with integer coefficients, for which integer solutions are sought. Diophantus also made advances in mathematical notation.

DIOPHANTI
ALEXANDRINI
ARITHMETICORVM
LIBRI SEX,
ET DE NVMERIS MVLTANGVLIS
LIBER VNVS.
Nunc primum Græcè & Latinè editi, atque absolutissimis Commentariis illustrati.
AVCTORE CLAVDIO GASPARE BACHETO MEZIRIACO SEBVSIANO, V.C.
LVTETIAE PARISIORVM,
Sumptibus SEBASTIANI CRAMOISY, via Iacobæa, sub Ciconiis.
M. DC. XXI.
CVM PRIVILEGIO REGIS.

Title page of the 1621 edition of Diophantus' *Arithmetica*, translated into Latin by Claude Gaspard Bachet de Méziriac.

Biography

Little is known about the life of Diophantus. He lived in Alexandria, Egypt, probably from between A.D. 200 and 214 to 284 or 298. Much of our knowledge of the life of Diophantus is derived from a 5th century Greek anthology of number games and strategy puzzles. One of the problems (sometimes called his epitaph) states:

> 'Here lies Diophantus,' the wonder behold.
>
> Through art algebraic, the stone tells how old:
>
> 'God gave him his boyhood one-sixth of his life,
>
> One twelfth more as youth while whiskers grew rife;
>
> And then yet one-seventh ere marriage begun;
>
> In five years there came a bouncing new son.
>
> Alas, the dear child of master and sage
>
> After attaining half the measure of his father's life chill fate took him. After consoling his fate by the science of numbers for four years, he ended his life.'

This puzzle implies that Diophantus lived to be 84 years old. However, the accuracy of the information cannot be independently confirmed.

In popular culture, this puzzle was the Puzzle No.142 in *Professor Layton and Pandora's Box* as one of the hardest solving puzzles in the game, which needed to be unlocked by solving other puzzles first.

Arithmetica

The Arithmetica is the major work of Diophantus and the most prominent work on algebra in Greek mathematics. It is a collection of problems giving numerical solutions of both determinate and indeterminate equations. Of the original thirteen books of which Arithmetica consisted only six have survived, though there are some who believe that four Arab books discovered in 1968 are also by Diophantus.[6] Some Diophantine problems from Arithmetica have been found in Arabic sources.

It should be mentioned here that Diophantus never used general methods in his solutions. Hermann Hankel, renowned German mathematician made the following remark regarding Diophantus.

"Our author (Diophantos) not the slightest trace of a general, comprehensive method is discernible; each problem calls for some special method which refuses to work even for the most closely related problems. For this reason it is difficult for the modern scholar to solve the 101st problem even after having studied 100 of Diophantos's solutions" [7]

History

Like many other Greek mathematical treatises, Diophantus was forgotten in Western Europe during the so-called Dark Ages, since the study of ancient Greek had greatly declined. The portion of the Greek *Arithmetica* that survived, however, was, like all ancient Greek texts transmitted to the early modern world, copied by, and thus known to, medieval Byzantine scholars. In addition, some portion of the *Arithmetica* probably survived in the Arab tradition (see above). In 1463 German mathematician Regiomontanus wrote:

> "No one has yet translated from the Greek into Latin the thirteen books of Diophantus, in which the very flower of the whole of arithmetic lies hidden . . . ."

Arithmetica was first translated from Greek into Latin by Bombelli in 1570, but the translation was never published. However, Bombelli borrowed many of the problems for his own book *Algebra*. The *editio princeps* of *Arithmetica* was published in 1575 by Xylander. The best known Latin translation of *Arithmetica* was made by Bachet in 1621 and became the first Latin edition that was widely available. Pierre de Fermat owned a copy, studied it, and made notes in the margins.

Margin writing by Fermat and Chortasmenos

The 1621 edition of Arithmetica by Bachet gained fame after Pierre de Fermat wrote his famous "Last Theorem" in the margins of his copy:

> "If an integer *n* is greater than 2, then has no solutions in non-zero integers *a*, *b*, and *c*. I have a truly marvelous proof of this proposition which this margin is too narrow to contain."

Fermat's proof was never found, and the problem of finding a proof for the theorem went unsolved for centuries. A proof was finally found in 1994 by Andrew Wiles after working on it for seven years. It is believed that Fermat did not actually have the proof he claimed to have. Although the original copy in which Fermat wrote this is lost today, Fermat's son edited the next edition of Diophantus, published in 1670. Even though the text is otherwise inferior to the 1621 edition, Fermat's annotations—including the "Last Theorem"—were printed in this version.

Fermat was not the first mathematician so moved to write in his own marginal notes to Diophantus; the Byzantine scholar John Chortasmenos (14th/15th C.) had written "Thy soul, Diophantus, be with Satan because of the difficulty of your theorems" next to the same problem.

Other works

Diophantus wrote several other books besides *Arithmetica*, but very few of them have survived.

The *Porisms*

Diophantus himself refers to a work which consists of a collection of lemmas called *The Porisms* (or *Porismata*), but this book is entirely lost. Some scholars think that *The porisms* may have actually been a section of *Arithmetica* that is now lost.

Although *The Porisms* is lost, we know three lemmas contained there, since Diophantus refers to them in the *Arithmetica*. One lemma states that the difference of the cubes of two rational numbers is equal to the sum of the cubes of two other rational numbers, i.e. given any a and b, with $a > b$, there exist c and d, all positive and rational, such that

Polygonal numbers and geometric elements

Diophantus is also known to have written on polygonal numbers, a topic of great interest to Pythagoras and Pythagoreans. Fragments of a book dealing with polygonal numbers are extant.

A book called *Preliminaries to the Geometric Elements* has been traditionally attributed to Hero of Alexandria. It has been studied recently by Wilbur Knorr, who suggested that the attribution to Hero is incorrect, and that the true author is Diophantus.[8]

Influence

Diophantus' work has had a large influence in history. Editions of Arithmetica exerted a profound influence on the development of algebra in Europe in the late sixteenth and through the seventeenth and eighteenth centuries. Diophantus and his works have also influenced Arab mathematics and were of great fame among Arab mathematicians. Diophantus' work created a foundation for work on algebra and in fact much of advanced mathematics is based on algebra. As far as we know Diophantus did not affect the lands of the Orient much and how much he affected India is a matter of debate.

The father of algebra?

Diophantus is often called "the father of algebra" because he contributed greatly to number theory, mathematical notation, and because Arithmetica contains the earliest known use of syncopated notation.[9] However, it seems that many of the methods for solving linear and quadratic equations used by Diophantus go back to Babylonian mathematics. For this, and other, reasons mathematical historian Kurt Vogel writes: "Diophantus was not, as he has often been called, the father of algebra. Nevertheless, his remarkable, if unsystematic, collection of indeterminate problems is a singular achievement that was not fully appreciated and further developed until much later."[10]

Diophantine analysis

Today Diophantine analysis is the area of study where integer (whole number) solutions are sought for equations, and Diophantine equations are polynomial equations with integer coefficients to which only integer solutions are sought. It is usually rather difficult to tell whether a given Diophantine equation is solvable. Most of the problems in Arithmetica lead to quadratic equations. Diophantus looked at 3 different types of quadratic equations: , , and . The reason why there were three cases to Diophantus, while today we have only one case, is that he did not have any notion for zero and he avoided negative coefficients by considering the given numbers to all be positive in each of the three cases above. Diophantus was always satisfied with a rational solution and did not require a whole number which means he accepted fractions as solutions to his problems. Diophantus considered negative or irrational square root solutions "useless", "meaningless", and even "absurd". To give one specific example, he calls the equation 'absurd' because it would lead to a negative value for x. One solution was all he looked for in a quadratic equation. There is no evidence that suggests Diophantus even realized that there could be two solutions to a quadratic equation. He also considered simultaneous quadratic equations.

Mathematical notation

Diophantus made important advances in mathematical notation. He was the first person to use algebraic notation and symbolism. Before him everyone wrote out equations completely. Diophantus introduced an algebraic symbolism that used an abridged notation for frequently occurring operations, and an abbreviation for the unknown and for the powers of the unknown. Mathematical historian Kurt Vogel states:

"The symbolism that Diophantus introduced for the first time, and undoubtedly devised himself, provided a short and readily comprehensible means of expressing an equation... Since an abbreviation is also employed for the word 'equals', Diophantus took a fundamental step from verbal algebra towards symbolic algebra."

Although Diophantus made important advances in symbolism, he still lacked the necessary notation to express more general methods. This caused his work to be more concerned with particular problems rather than general situations. Some of the limitations of Diophantus' notation are that he only had notation for one unknown and, when problems involved more than a single unknown, Diophantus was reduced to expressing "first unknown", "second unknown", etc. in words. He also lacked a symbol for a general number n. Where we would write , Diophantus has to resort to constructions like : ... a sixfold number increased by twelve, which is divided by the difference by which the square of the number exceeds three.

Algebra still had a long way to go before very general problems could be written down and solved succinctly.

See also

- Erdős–Diophantine graph
- Diophantus II.VIII
- Polynomial Diophantine equation

Notes

[1] Research Machines plc. (2004). *The Hutchinson dictionary of scientific biography*. Abingdon, Oxon: Helicon Publishing. p. 312. "**Diophantus (lived c.A.D. 270-280)** Greek mathematician who, in solving linear mathematical problems, developed an early form of algebra."

[2] Boyer, Carl B. (1991). "Revival and Decline of Greek Mathematics". *A History of Mathematics* (Second ed.). John Wiley & Sons, Inc. p. 178. ISBN 0-471-54397-7. "At the beginning of this period, also known as the Later Alexandrian Age, we find the leading Greek algebraist, Diophantus of Alexandria, and toward its close there appeared the last significant Greek geometer, Pappus of Alexandria."

[3] Cooke, Roger (1997). "The Nature of Mathematics". *The History of Mathematics: A Brief Course*. Wiley-Interscience. p. 7. ISBN 0-471-18082-3. "Some enlargement in the sphere in which symbols were used occurred in the writings of the third-century Greek mathematician Diophantus of Alexandria, but the same defect was present as in the case of Akkadians."

[4] Victor J. Katz (1998). *A History of Mathematics: An Introduction*, p. 184. Addison Wesley, ISBN 0-321-01618-1.

"But what we really want to know is to what extent the Alexandrian mathematicians of the period from the first to the fifth centuries C.E. were Greek. Certainly, all of them wrote in Greek and were part of the Greek intellectual community of Alexandria. And most modern studies conclude that the Greek community coexisted [...] So should we assume that Ptolemy and Diophantus, Pappus and Hypatia were ethnically Greek, that their ancestors had come from Greece at some point in the past but had remained effectively isolated from the Egyptians? It is, of course, impossible to answer this question definitively. But research in papyri dating from the early centuries of the common era demonstrates that a significant amount of intermarriage took place between the Greek and Egyptian communities [...] And it is known that Greek marriage contracts increasingly came to resemble Egyptian ones. In addition, even from the founding of Alexandria, small numbers of Egyptians were admitted to the privaleged classes in the city to fulfill numerous civic roles. Of course, it was essential in such cases for the Egyptians to become "Hellenized," to adopt Greek habits and the Greek language. Given that the Alexandrian mathematicians mentioned here were active several hundred years after the founding of the city, it would seem at least equally possible that they were ethnically Egyptian as that they remained ethnically Greek. In any case, it is unreasonable to portray them with purely European features when no physical descriptions exist."

[5] Katz, Mikhail G.; Schaps, David; Shnider, Steve (2013), "Almost Equal: The Method of Adequality from Diophantus to Fermat and Beyond", *Perspectives on Science* **21** (3), arXiv:1210.7750

[6] J. Sesiano (1982). *Books IV to VII of Diophantus'* Arithmetica *in the Arabic Translation Attributed to Qusta ibn Luqa*. New York/Heidelberg/Berlin: Springer-Verlag. p. 502.

[7] Hankel H., "Geschichte der mathematic im altertum und mittelalter, Leipzig, 1874. (translated to English by Ulrich Lirecht in Chinese Mathematics in the thirteenth century, Dover publications, New York, 1973.

[8] Knorr, Wilbur: Arithmêtike stoicheiôsis: On Diophantus and Hero of Alexandria, in: Historia Matematica, New York, 1993, Vol.20, No.2, 180-192

[9] Carl B. Boyer, A History of Mathematics, Second Edition (Wiley, 1991), page 228

[10] Harald Kittel, *Übersetzung: ein internationales Handbuch zur Übersetzungsforschung*, Volume 2 p. 1123, 1124

References

- Allard, A. "Les scolies aux arithmétiques de Diophante d'Alexandrie dans le Matritensis Bibl.Nat.4678 et les Vatican Gr.191 et 304" *Byzantion* 53. Brussels, 1983: 682-710.
- Christianidis, J. "Maxime Planude sur le sens du terme diophantien "plasmatikon"", *Historia Scientiarum*, 6 (1996)37-41.
- Christianidis, J. "Une interpretation byzantine de Diophante", *Historia Mathematica*, 25 (1998) 22-28.
- Heath, Sir Thomas, *Diophantos of Alexandria: A Study in the History of Greek Algebra*, Cambridge: Cambridge University Press, 1885, 1910.
- Robinson, D. C. and Luke Hodgkin. *History of Mathematics*, King's College London, 2003.
- Sesiano, Jacques. *Books IV to VII of Diophantus' Arithmetica in the Arabic translation attributed to Qusṭā ibn Lūqā*, Heidelberg: Springer-Verlag, 1982. ISBN 0-387-90690-8.
- Tannery, P. L. *Diophanti Alexandrini Opera omnia: cum Graecis commentariis*, Lipsiae: In aedibus B.G. Teubneri, 1893-1895.
- Ver Eecke, P. *Diophante d'Alexandrie: Les Six Livres Arithmétiques et le Livre des Nombres Polygones*, Bruges: Desclée, De Brouwer, 1921.

Further reading

- Heath, Sir Thomas (1981). *A history of Greek mathematics* (http://www.archive.org/details/ahistorygreekma00heatgoog). **2**. Cambridge University Press: Cambridge.
- Vogel, Kurt (1970). "Diophantus of Alexandria". *Dictionary of Scientific Biography*. **4**. New York: Scribner.

External links

- O'Connor, John J.; Robertson, Edmund F., "Diophantus" (http://www-history.mcs.st-andrews.ac.uk/Biographies/Diophantus.html), *MacTutor History of Mathematics archive*, University of St Andrews.
- Diophantus's Riddle (http://mathworld.wolfram.com/DiophantussRiddle.html) Diophantus' epitaph, by E. Weisstein
- Norbert Schappacher (2005). Diophantus of Alexandria : a Text and its History (http://www-irma.u-strasbg.fr/~schappa/NSch/Publications_files/Dioph.pdf).
- Tannery's edition of the Works of Diophantus, now in the public domain (http://www.wilbourhall.org) (Classical Greek)
- Review of Sesiano's Diophantus (http://www.math.uu.nl/people/hogend/sesiano.html) Review of J. Sesiano, Books IV to VII of Diophantus' Arithmetica, by Jan P. Hogendijk

- Latin translation from 1575 by [[Wilhelm Xylander (http://echo.mpiwg-berlin.mpg.de/ECHOdocuViewfull?mode=imagepath&url=/mpiwg/online/permanent/library/W770Y3H9/pageimg&viewMode=images)]]

Euclid

Euclid

Euclid in Raphael's *School of Athens*

Born	Unknown
Died	Unknown
Residence	Alexandria, Egypt
Fields	Mathematics
Known for	Euclidean geometry Euclid's *Elements*

Euclid (pron.: /ˈjuːklɪd/ ***EWK**-lid*; Ancient Greek: Εὐκλείδης *Eukleidēs*), fl. 300 BC, also known as **Euclid of Alexandria**, was a Greek mathematician, often referred to as the "Father of Geometry". He was active in Alexandria during the reign of Ptolemy I (323–283 BC). His *Elements* is one of the most influential works in the history of mathematics, serving as the main textbook for teaching mathematics (especially geometry) from the time of its publication until the late 19th or early 20th century.[1] [2] [3] In the *Elements*, Euclid deduced the principles of what is now called Euclidean geometry from a small set of axioms. Euclid also wrote works on perspective, conic sections, spherical geometry, number theory and rigor.

"Euclid" is the anglicized version of the Greek name Εὐκλείδης, meaning "Good Glory".[4]

Life

Little is known about Euclid's life, as there are only a handful of references to him. The date and place of Euclid's birth and the date and circumstances of his death are unknown, and only roughly estimated in proximity to contemporary figures mentioned in references. No likeness or description of Euclid's physical appearance made during his lifetime survived antiquity. Therefore, Euclid's depiction in works of art is the product of the artist's imagination.

The few historical references to Euclid were written centuries after he lived, by Proclus and Pappus of Alexandria.[5] Proclus introduces Euclid only briefly in his fifth-century *Commentary on the Elements*, as the author of *Elements*, that he was mentioned by Archimedes, and that when King Ptolemy asked if there was a shorter path to learning geometry than Euclid's *Elements*, "Euclid replied there is no royal road to geometry."[6] Although the purported citation of Euclid by Archimedes has been judged to be an interpolation by later editors of his works, it is still believed that Euclid wrote his works before those of Archimedes.[7] [8] [9] In addition, the "royal road" anecdote is questionable since it is similar to a story told about Menaechmus and Alexander the Great.[10] In the only other key reference to Euclid, Pappus briefly mentioned in the fourth century that Apollonius "spent a very long time with the pupils of Euclid at Alexandria, and it was thus that he acquired such a scientific habit of thought."[11] It is further believed that Euclid may have studied at Plato's Academy in Athens.

Elements

Although many of the results in *Elements* originated with earlier mathematicians, one of Euclid's accomplishments was to present them in a single, logically coherent framework, making it easy to use and easy to reference, including a system of rigorous mathematical proofs that remains the basis of mathematics 23 centuries later.[12]

There is no mention of Euclid in the earliest remaining copies of the *Elements*, and most of the copies say they are "from the edition of Theon" or the "lectures of Theon",[13] while the text considered to be primary, held by the Vatican, mentions no author. The only reference that historians rely on of Euclid having written the *Elements* was from Proclus, who briefly in his *Commentary on the Elements* ascribes Euclid as its author.

Although best known for its geometric results, the *Elements* also includes number theory. It considers the connection between perfect numbers and Mersenne primes, the infinitude of prime numbers, Euclid's lemma on factorization (which leads to the fundamental theorem of arithmetic on uniqueness of prime factorizations), and the Euclidean algorithm for finding the greatest common divisor of two numbers.

The geometrical system described in the *Elements* was long known simply as *geometry*, and was considered to be the only geometry possible. Today, however, that system is often referred to as *Euclidean geometry* to distinguish it from other so-called *non-Euclidean geometries* that mathematicians discovered in the 19th century.

Other works

In addition to the *Elements*, at least five works of Euclid have survived to the present day. They follow the same logical structure as *Elements*, with definitions and proved propositions.

- *Data* deals with the nature and implications of "given" information in geometrical problems; the subject matter is closely related to the first four books of the *Elements*.
- *On Divisions of Figures*, which survives only partially in Arabic translation, concerns the division of geometrical figures into two or more equal parts or into parts in given ratios. It is similar to a third century AD work by Heron of Alexandria.
- *Catoptrics*, which concerns the mathematical theory of mirrors, particularly the images formed in plane and spherical concave mirrors. The attribution is held to be anachronistic however by J J O'Connor and E F Robertson who name Theon of Alexandria as a more likely author.[14]
- *Phaenomena*, a treatise on spherical astronomy, survives in Greek; it is quite similar to *On the Moving Sphere* by Autolycus of Pitane, who flourished around 310 BC.

- *Optics* is the earliest surviving Greek treatise on perspective. In its definitions Euclid follows the Platonic tradition that vision is caused by discrete rays which emanate from the eye. One important definition is the fourth: "Things seen under a greater angle appear greater, and those under a lesser angle less, while those under equal angles appear equal." In the 36 propositions that follow, Euclid relates the apparent size of an object to its distance from the eye and investigates the apparent shapes of cylinders and cones when viewed from different angles. Proposition 45 is interesting, proving that for any two unequal magnitudes, there is a point from which the two appear equal. Pappus believed these results to be important in astronomy and included Euclid's *Optics*, along with his *Phaenomena*, in the *Little Astronomy*, a compendium of smaller works to be studied before the *Syntaxis* (*Almagest*) of Claudius Ptolemy.

Statue of Euclid in the Oxford University Museum of Natural History

Other works are credibly attributed to Euclid, but have been lost.

- *Conics* was a work on conic sections that was later extended by Apollonius of Perga into his famous work on the subject. It is likely that the first four books of Apollonius's work come directly from Euclid. According to Pappus, "Apollonius, having completed Euclid's four books of conics and added four others, handed down eight volumes of conics." The Conics of Apollonius quickly supplanted the former work, and by the time of Pappus, Euclid's work was already lost.
- *Porisms* might have been an outgrowth of Euclid's work with conic sections, but the exact meaning of the title is controversial.
- *Pseudaria*, or *Book of Fallacies*, was an elementary text about errors in reasoning.
- *Surface Loci* concerned either loci (sets of points) on surfaces or loci which were themselves surfaces; under the latter interpretation, it has been hypothesized that the work might have dealt with quadric surfaces.
- Several works on mechanics are attributed to Euclid by Arabic sources. *On the Heavy and the Light* contains, in nine definitions and five propositions, Aristotelian notions of moving bodies and the concept of specific gravity. *On the Balance* treats the theory of the lever in a similarly Euclidean manner, containing one definition, two axioms, and four propositions. A third fragment, on the circles described by the ends of a moving lever, contains four propositions. These three works complement each other in such a way that it has been suggested that they are remnants of a single treatise on mechanics written by Euclid.

See also

- Axiomatic method
- Euclid's orchard
- Euclidean algorithm
- Euclidean geometry
- Euclidean relation
- Extended Euclidean algorithm
- List of topics named after Euclid
- *Papyrus Oxyrhynchus 29*

Notes

[1] Ball, pp. 50–62.

[2] Boyer, pp. 100–19.

[3] Macardle, et al. (2008). *Scientists: Extraordinary People Who Altered the Course of History.* New York: Metro Books. g. 12.

[4] etymonline.com (http://www.etymonline.com/index.php?allowed_in_frame=0&search=Euclid&searchmode=none) Retrieved 2011-12-04

[5] Joyce, David. *Euclid.* Clark University Department of Mathematics and Computer Science. (http://aleph0.clarku.edu/~djoyce/java/elements/Euclid.html)

[6] Proclus, p. 57 (http://books.google.com/books?id=JZEHj2fEmqAC&lpg=PP1&pg=PA57#v=onepage&q&f=false)

[7] Proclus, p. XXX (http://books.google.com/books?id=JZEHj2fEmqAC&pg=PA88&vq=euclid&dq=proclus+commentary+elements&source=gbs_search_s&cad=0#PPR30,M1)

[8] *Euclid of Alexandria* (http://www-history.mcs.st-and.ac.uk/Biographies/Euclid.html)

[9] The MacTutor History of Mathematics archive.

[10] Boyer, p. 96.

[11] Heath (1956), p. 2.

[12] Struik p. 51 ("their logical structure has influenced scientific thinking perhaps more than any other text in the world").

[13] Heath (1981), p. 360.

[14] Theon article at their institution (http://www-history.mcs.st-andrews.ac.uk/Biographies/Theon.html)

References

- "Euclid (Greek mathematician)" (http://www.britannica.com/EBchecked/topic/194880/Euclid). Encyclopædia Britannica, Inc. 2008. Retrieved 2008-04-18.
- Artmann, Benno (1999). *Euclid: The Creation of Mathematics*. New York: Springer. ISBN 0-387-98423-2.
- Ball, W.W. Rouse (1960) [1908]. *A Short Account of the History of Mathematics* (4th ed.). Dover Publications. pp. 50–62. ISBN 0-486-20630-0.
- Boyer, Carl B. (1991). *A History of Mathematics* (2nd ed.). John Wiley & Sons, Inc.. ISBN 0-471-54397-7.
- Heath, Thomas (ed.) (1956) [1908]. *The Thirteen Books of Euclid's Elements*. **1**. Dover Publications. ISBN 0-486-60088-2.
- Heath, Thomas L. (1908), " Euclid and the Traditions About Him (http://perseus.mpiwg-berlin.mpg.de/cgi-bin/ptext?lookup=Euc.+1)", in Euclid, *Elements* (Thomas L. Heath, ed. 1908), **1**:1–6, at Perseus Digital Library (http://perseus.mpiwg-berlin.mpg.de/).
- Heath, Thomas L. (1981). *A History of Greek Mathematics*, 2 Vols. New York: Dover Publications. ISBN 0-486-24073-8 / ISBN 0-486-24074-6.
- Kline, Morris (1980). *Mathematics: The Loss of Certainty*. Oxford: Oxford University Press. ISBN 0-19-502754-X.
- O'Connor, John J.; Robertson, Edmund F., "Euclid of Alexandria" (http://www-history.mcs.st-andrews.ac.uk/Biographies/Euclid.html), *MacTutor History of Mathematics archive*, University of St Andrews.
- Proclus, *A commentary on the First Book of Euclid's Elements*, translated by Glenn Raymond Morrow, Princeton University Press, 1992. ISBN 978-0-691-02090-7.
- Struik, Dirk J. (1967). *A Concise History of Mathematics*. Dover Publications. ISBN 486-60255-9.

Further reading

- DeLacy, Estelle Allen (1963). *Euclid and Geometry*. New York: Franklin Watts.
- Knorr, Wilbur Richard (1975). *The Evolution of the Euclidean Elements: A Study of the Theory of Incommensurable Magnitudes and Its Significance for Early Greek Geometry*. Dordrecht, Holland: D. Reidel. ISBN 90-277-0509-7.
- Mueller, Ian (1981). *Philosophy of Mathematics and Deductive Structure in Euclid's Elements*. Cambridge, MA: MIT Press. ISBN 0-262-13163-3.
- Reid, Constance (1963). *A Long Way from Euclid*. New York: Crowell.
- Szabó, Árpád (1978). *The Beginnings of Greek Mathematics*. A.M. Ungar, trans. Dordrecht, Holland: D. Reidel. ISBN 90-277-0819-3.

External links

- Euclid's Elements (http://aleph0.clarku.edu/~djoyce/java/elements/elements.html), All thirteen books, with interactive diagrams using Java. Clark University
- Euclid's Elements (http://farside.ph.utexas.edu/euclid.html), with the original Greek and an English translation on facing pages (includes PDF version for printing). University of Texas.
- Euclid's Elements, books I-VI (http://www.gutenberg.org/files/21076/21076-pdf), in English pdf, in a Project Gutenberg Victorian textbook edition with diagrams.
- Euclid's Elements (http://euclides.org), All thirteen books, in several languages as Spanish, Catalan, English, German, Portuguese, Arabic, Italian, Russian and Chinese.
- *Elementa Geometriae* (http://www.rarebookroom.org/Control/eucgeo/index.html) 1482, Venice. From Rare Book Room.
- *Elementa* (http://www.rarebookroom.org/Control/eucmsd/index.html) 888 AD, Byzantine. From Rare Book Room.
- Euclid biography by Charlene Douglass (http://www.mathopenref.com/euclid.html) With extensive bibliography.
- Texts on Ancient Mathematics and Mathematical Astronomy (http://www.wilbourhall.org) PDF scans (Note: many are very large files). Includes editions and translations of Euclid's *Elements*, *Data*, and *Optica*, Proclus's *Commentary on Euclid*, and other historical sources.

Eudoxus_of_Cnidus

Eudoxus of Cnidus (410 or 408 BC – 355 or 347 BC) was a Greek astronomer, mathematician, scholar and student of Plato. Since all his own works are lost, knowledge of him is obtained from secondary sources, such as Aratus's poem on astronomy. Theodosius of Bithynia's important work, Sphaerics, may be based on a work of Eudoxus.

Life

His name Eudoxus means "honored" or "of good repute" (in Greek Εὔδοξος, from *eu* "good" and *doxa* "opinion, belief, fame"). It is analogous to the Latin name Benedictus.

Eudoxus's father Aeschines of Cnidus loved to watch stars at night. Eudoxus first travelled to Tarentum to study with Archytas, from whom he learned mathematics. While in Italy, Eudoxus visited Sicily, where he studied medicine with Philiston.

Around 387 BC, at the age of 23, he traveled with the physician Theomedon, who according to Diogenes Laërtius some believed was his lover,[1] to Athens to study with the followers of Socrates. He eventually became the pupil of Plato, with whom he studied for several months, but due to a disagreement they had a falling out. Eudoxus was quite poor and could only afford an apartment at the Piraeus. To attend Plato's lectures, he walked the seven miles (11 km) each direction, each day. Due to his poverty, his friends raised funds sufficient to send him to Heliopolis, Egypt to pursue his study of astronomy and mathematics. He lived there for 16 months. From Egypt, he then traveled north to Cyzicus, located on the south shore of the Sea of Marmara, the Propontis. He traveled south to the court of Mausolus. During his travels he gathered many students of his own.

Around 368 BC, Eudoxus returned to Athens with his students. According to some sources, around 367 he assumed headship of the Academy during Plato's period in Syracuse, and taught Aristotle. He eventually returned to his native Cnidus, where he served in the city assembly. While in Cnidus, he built an observatory and continued writing and lecturing on theology, astronomy and meteorology. He had one son, Aristagoras, and three daughters, Actis, Philtis and Delphis.

In mathematical astronomy, his fame is due to the introduction of the astronomical globe, and his early contributions to understanding the movement of the planets.

His work on proportions shows tremendous insight into numbers; it allows rigorous treatment of continuous quantities and not just whole numbers or even rational numbers. When it was revived by Tartaglia and others in the 16th century, it became the basis for quantitative work in science for a century, until it was replaced by the algebraic methods of Descartes.

Craters on Mars and the Moon are named in his honor. An algebraic curve (the Kampyle of Eudoxus) is also named after him

$$a^2x^4 = b^4(x^2 + y^2).$$

Mathematics

Eudoxus is considered by some to be the greatest of classical Greek mathematicians, and in all antiquity, second only to Archimedes. He rigorously developed Antiphon's method of exhaustion, a precursor to the integral calculus which was also used in a masterly way by Archimedes in the following century. In applying the method, Eudoxus proved such mathematical statements as: areas of circles are to one another as the squares of their radii, volumes of spheres are to one another as the cubes of their radii, the volume of a pyramid is one-third the volume of a prism with the same base and altitude, and the volume of a cone is one-third that of the corresponding cylinder.[2]

Eudoxus introduced the idea of non-quantified mathematical magnitude to describe and work with continuous geometrical entities such as lines, angles, areas and volumes, thereby avoiding the use of irrational numbers. In doing so, he reversed a Pythagorean emphasis on number and arithmetic, focusing instead on geometrical concepts as the basis of rigorous mathematics. Some Pythagoreans, such as Eudoxus' teacher Archytas, had believed that only arithmetic could provide a basis for proofs. Induced by the need to understand and operate with incommensurable quantities, Eudoxus established what may have been the first deductive organization of mathematics on the basis of explicit axioms. The change in focus by Eudoxus stimulated a divide in mathematics which lasted two thousand years. In combination with a Greek intellectual attitude unconcerned with practical problems, there followed a significant retreat from the development of techniques in arithmetic and algebra.[3]

The Pythagoreans had discovered that the diagonal of a square does not have a common unit of measurement with the sides of the square; this is the famous discovery that the square root of 2 cannot be expressed as the ratio of two integers. This discovery had heralded the existence of incommensurable quantities beyond the integers and rational fractions, but at the same time it threw into question the idea of measurement and calculations in geometry as a whole. For example, Euclid provides an elaborate proof of the Pythagorean theorem (*Elements* I.47), by using addition of areas and only much later (*Elements* VI.31) a simpler proof from similar triangles, which relies on ratios of line segments.

Ancient Greek mathematicians calculated not with quantities and equations as we do today, but instead they used proportionalities to express the relationship between quantities. Thus the ratio of two similar quantities was not just a numerical value, as we think of it today; the ratio of two similar quantities was a primitive relationship between them.

Eudoxus was able to restore confidence in the use of proportionalities by providing an astounding definition for the meaning of the equality between two ratios. This definition of proportion forms the subject of Euclid's Book V.

In Definition 5 of Euclid's Book V we read:

> Magnitudes are said to be in the same ratio, the first to the second and the third to the fourth when, if any equimultiples whatever be taken of the first and third, and any equimultiples whatever of the second and fourth, the former equimultiples alike exceed, are alike equal to, or alike fall short of, the latter equimultiples respectively taken in corresponding order.

Let us clarify it by using modern-day notation. If we take four quantities: a, b, c, and d, then the first and second have a ratio ; similarly the third and fourth have a ratio .

Now to say that we do the following: For any two arbitrary integers, m and n, form the equimultiples $m{\cdot}a$ and $m{\cdot}c$ of the first and third; likewise form the equimultiples $n{\cdot}b$ and $n{\cdot}d$ of the second and fourth.

If it happens that $m{\cdot}a > n{\cdot}b$, then we must also have $m{\cdot}c > n{\cdot}d$. If it happens that $m{\cdot}a = n{\cdot}b$, then we must also have $m{\cdot}c = n{\cdot}d$. Finally, if it happens that $m{\cdot}a < n{\cdot}b$, then we must also have $m{\cdot}c < n{\cdot}d$.

Notice that the definition depends on comparing the similar quantities $m{\cdot}a$ and $n{\cdot}b$, and the similar quantities $m{\cdot}c$ and $n{\cdot}d$, and does not depend on the existence of a common unit of measuring these quantities.

The complexity of the definition reflects the deep conceptual and methodological innovation involved. It brings to mind the famous fifth postulate of Euclid concerning parallels, which is more extensive and complicated in its wording than the other postulates.

The Eudoxian definition of proportionality uses the quantifier, "for every ..." to harness the infinite and the infinitesimal, just as do the modern epsilon-delta definitions of limit and continuity.

Additionally, the Archimedean property stated as definition 4 of Euclid's book V is originally due not to Archimedes but to Eudoxus.[4]

Astronomy

In ancient Greece, astronomy was a branch of mathematics; astronomers sought to create geometrical models that could imitate the appearances of celestial motions. Identifying the astronomical work of Eudoxus as a separate category is therefore a modern convenience. Some of Eudoxus' astronomical texts whose names have survived include:

- *Disappearances of the Sun*, possibly on eclipses
- *Oktaeteris* (Ὀκταετηρίς), on an eight-year lunisolar cycle of the calendar
- *Phaenomena* (Φαινόμενα) and *Entropon* (Ἔντροπον), on spherical astronomy, probably based on observations made by Eudoxus in Egypt and Cnidus
- *On Speeds*, on planetary motions

We are fairly well informed about the contents of *Phaenomena*, for Eudoxus' prose text was the basis for a poem of the same name by Aratus. Hipparchus quoted from the text of Eudoxus in his commentary on Aratus.

Eudoxan planetary models

A general idea of the content of *On Speeds* can be gleaned from Aristotle's *Metaphysics* XII, 8, and a commentary by Simplicius of Cilicia (6th century CE) on *De caelo*, another work by Aristotle. According to a story reported by Simplicius, Plato posed a question for Greek astronomers: "By the assumption of what uniform and orderly motions can the apparent motions of the planets be accounted for?" (quoted in Lloyd 1970, p. 84). Plato proposed that the seemingly chaotic wandering motions of the planets could be explained by combinations of uniform circular motions centered on a spherical Earth, apparently a novel idea in the 4th century.

In most modern reconstructions of the Eudoxan model, the Moon is assigned three spheres:

- The outermost rotates westward once in 24 hours, explaining rising and setting.
- The second rotates eastward once in a month, explaining the monthly motion of the Moon through the zodiac.
- The third also completes its revolution in a month, but its axis is tilted at a slightly different angle, explaining motion in latitude (deviation from the ecliptic), and the motion of the lunar nodes.

The Sun is also assigned three spheres. The second completes its motion in a year instead of a month. The inclusion of a third sphere implies that Eudoxus mistakenly believed that the Sun had motion in latitude.

The five visible planets (Venus, Mercury, Mars, Jupiter, and Saturn) are assigned four spheres each:

- The outermost explains the daily motion.
- The second explains the planet's motion through the zodiac.
- The third and fourth together explain retrogradation, when a planet appears to slow down, then briefly reverse its motion through the zodiac. By inclining the axes of the two spheres with respect to each other, and rotating them in opposite directions but with equal periods, Eudoxus could make a point on the inner sphere trace out a figure-eight shape, or hippopede.

Importance of Eudoxan system

Callippus, a Greek astronomer of the 4th century, added seven spheres to Eudoxus' original 27 (in addition to the planetary spheres, Eudoxus included a sphere for the fixed stars). Aristotle described both systems, but insisted on adding "unrolling" spheres between each set of spheres to cancel the motions of the outer set. Aristotle was concerned about the physical nature of the system; without unrollers, the outer motions would be transferred to the inner planets.

A major flaw in the Eudoxan system is its inability to explain changes in the brightness of planets as seen from Earth. Because the spheres are concentric, planets will always remain at the same distance from Earth. This problem was pointed out in Antiquity by Autolycus of Pitane. Astronomers responded by introducing the deferent and epicycle, which caused a planet to vary its distance. However, Eudoxus' importance to Greek astronomy is considerable, as he was the first to attempt a mathematical explanation of the planets.

Ethics

Aristotle, in The Nicomachean Ethics[5] attributes to Eudoxus an argument in favor of hedonism, that is, that pleasure is the ultimate good that activity strives for. According to Aristotle, Eudoxus put forward the following arguments for this position:

1. All things, rational and irrational, aim at pleasure; things aim at what they believe to be good; a good indication of what the chief good is would be the thing that most things aim at.
2. Similarly, pleasure's opposite – pain – is universally avoided, which provides additional support for the idea that pleasure is universally considered good.
3. People don't seek pleasure as a means to something else, but as an end in its own right.
4. Any other good that you can think of would be better if pleasure were added to it, and it is only by good that good can be increased.
5. Of all of the things that are good, happiness is peculiar for not being praised, which may show that it is the crowning good.[6]

See also

- Incommensurable magnitudes

References

- Evans, James (1998). *The History and Practice of Ancient Astronomy*. Oxford University Press. ISBN 0-19-509539-1. OCLC 185509676.
- Huxley, GL (1980). *Eudoxus of Cnidus* p. 465-7 in: *the Dictionary of Scientific Biography, volume 4.*
- Lloyd, GER (1970). *Early Greek Science: Thales to Aristotle*. W.W. Norton.

Notes

[1] Diogenes Laertius; VIII.87
[2] Morris Kline, *Mathematical Thought from Ancient to Modern Times* Oxford University Press, 1972 pp. 48-50
[3] ibid
[4] Knopp, Konrad (1951). *Theory and Application of Infinite Series* (English 2nd ed.). London and Glasgow: Blackie & Son, Ltd.. p. 7.
[5] largely in book ten
[6] this particular argument is referenced in book one

Further reading

- De Santillana, G. (1968). "Eudoxus and Plato: A Study in Chronology". *Reflections on Men and Ideas*. Cambridge, MA: MIT Press.
- Huxley, G. L. (1963). "Eudoxian Topics". *Greek, Roman, and Byzantine Studies* **4**: 83–96.
- Knorr, Wilbur Richard (1986). *The Ancient tradition of geometric problems*. Boston: Birkhäuser. ISBN 0-8176-3148-8.
- Knorr, Wilbur Richard (1978). "Archimedes and the Pre-Euclidean Proportion Theory". *Archives Intemationales d'histoire des Sciences* **28**: 183–244.
- Neugebauer, O. (1975). *A history of ancient mathematical astronomy*. Berlin: Springer-Verlag. ISBN 0-387-06995-X.
- Van der Waerden, B. L. (1988). *Science Awakening* (5th ed.). Leiden: Noordhoff.

External links

- Working model and complete explanation of the Eudoxus's Spheres (http://www.youtube.com/watch?v=_SFzDYSqR_4)
- Dennis Duke, "Statistical dating of the Phaenomena of Eudoxus", *DIO*, volume 15 (http://www.dioi.org/vols/wf0.pdf) *see pages 7 to 23*
- Diogenes Laërtius, *Life of Eudoxus*, translated by Robert Drew Hicks (1925). Wikisource
- Eudoxus of Cnidus (http://www.britannica.com/EBchecked/topic/195005/Eudoxus-of-Cnidus) Britannica.com
- Eudoxus of Cnidus (http://www.math.tamu.edu/~don.allen/history/eudoxus/eudoxus.html) Donald Allen, Professor, Texas A&M University
- Eudoxos of Knidos (Eudoxus of Cnidus): astronomy and homocentric spheres (http://www.calstatela.edu/faculty/hmendel/Ancient Mathematics/Eudoxus/Astronomy/EudoxusHomocentricSpheres.htm) Henry Mendell, Cal State U, LA
- Herodotus Project: Extensive B+W photo essay of Cnidus (http://www.losttrails.com/pages/Hproject/Caria/Cnidus/Cnidus.html)
- Models of Planetary Motion—Eudoxus (http://faculty.fullerton.edu/cmcconnell/Planets.html#3), Craig McConnell, Ph.D., Cal State, Fullerton
- O'Connor, John J.; Robertson, Edmund F., "Eudoxus of Cnidus" (http://www-history.mcs.st-andrews.ac.uk/Biographies/Eudoxus.html), *MacTutor History of Mathematics archive*, University of St Andrews.
- The Universe According to Eudoxus (http://hsci.cas.ou.edu/images/applets/hippopede.html) (Java applet)

Hipparchus

Hipparchus	
 Hipparchus from "The School of Athens" by Raphael	
Born	c. 190 B.C.Nicaea
Died	c. 120 B.C.Rhodes
Occupation	astronomer, mathematician, geographer
Influenced	Ptolemy

Hipparchus /hɪˈpɑːrkəs/ **of Nicaea**, or more correctly **Hipparchos** (Greek: Ἵππαρχος, *Hipparkhos*; c. 190 BC – c. 120 BC), was a Greek astronomer, geographer, and mathematician of the Hellenistic period. He is considered the founder of trigonometry[1] but is most famous for his incidental discovery of precession of the equinoxes.[2]

Hipparchus was born in Nicaea, Bithynia (now Iznik, Turkey), and probably died on the island of Rhodes. He is known to have been a working astronomer at least from 162 to 127 BC.[3] Hipparchus is considered the greatest ancient astronomical observer and, by some, the greatest overall astronomer of antiquity. He was the first whose quantitative and accurate models for the motion of the Sun and Moon survive. For this he certainly made use of the observations and perhaps the mathematical techniques accumulated over centuries by the Chaldeans from Babylonia. He developed trigonometry and constructed trigonometric tables, and he solved several problems of spherical trigonometry. With his solar and lunar theories and his trigonometry, he may have been the first to develop a reliable method to predict solar eclipses. His other reputed achievements include the discovery of Earth's precession, the compilation of the first comprehensive star catalog of the western world, and possibly the invention of the astrolabe, also of the armillary sphere, which he used during the creation of much of the star catalogue. It would be three centuries before Claudius Ptolemaeus' synthesis of astronomy would supersede the work of Hipparchus; it is heavily dependent on it in many areas.

Life and work

Relatively little of Hipparchus' direct work survives into modern times. Although he wrote at least fourteen books, only his commentary on the popular astronomical poem by Aratus was preserved by later copyists. Most of what is known about Hipparchus comes from Ptolemy's (2nd century) *Almagest*, with additional references to him by Pappus of Alexandria and Theon of Alexandria (c. 4th century AD) in their commentaries on the *Almagest*; from Strabo's *Geographia* ("Geography"), and from Pliny the Elder's *Naturalis historia* ("Natural history") (1st century AD).[4]

There is a strong tradition that Hipparchus was born in Nicaea (Greek *Νίκαια*), in the ancient district of Bithynia (modern-day Iznik in province Bursa), in what today is the country Turkey.

The exact dates of his life are not known, but Ptolemy attributes to him astronomical observations in the period from 147 BC to 127 BC, and some of these are stated as made in Rhodes; earlier observations since 162 BC might also have been made by him. His birth date (c. 190 BC) was calculated by Delambre based on clues in his work. Hipparchus must have lived some time after 127 BC because he analyzed and published his observations from that year. Hipparchus obtained information from Alexandria as well as Babylon, but it is not known when or if he visited these places. He is believed to have died on the island of Rhodes, where he seems to have spent most of his later life.

It is not known what Hipparchus' economic means were nor how he supported his scientific activities. His appearance is likewise unknown: there are no contemporary portraits. In the 2nd and 3rd centuries coins were made in his honour in Bithynia that bear his name and show him with a globe; this supports the tradition that he was born there.

Hipparchus is thought to be the first to calculate a heliocentric system, but he abandoned his work because the calculations showed the orbits were not perfectly circular as believed to be mandatory by the science of the time. As an astronomer of antiquity his influence, supported by Aristotle, held sway for nearly 2000 years, until the heliocentric model of Copernicus.

Hipparchus' only preserved work is *Τῶν Ἀράτου καὶ Εὐδόξου φαινομένων ἐξήγησις* ("Commentary on the Phaenomena of Eudoxus and Aratus"). This is a highly critical commentary in the form of two books on a popular poem by Aratus based on the work by Eudoxus.[5] Hipparchus also made a list of his major works, which apparently mentioned about fourteen books, but which is only known from references by later authors. His famous star catalog was incorporated into the one by Ptolemy, and may be almost perfectly reconstructed by subtraction of two and two thirds degrees from the longitudes of Ptolemy's stars. The first trigonometric table was apparently compiled by Hipparchus, who is now consequently known as "the father of trigonometry".

Modern speculation

Hipparchus was in the international news in 2005, when it was again proposed (as in 1898) that the data on the celestial globe of Hipparchus or in his star catalog may have been preserved in the only surviving large ancient celestial globe which depicts the constellations with moderate accuracy, the globe carried by the Farnese Atlas. There are a variety of mis-steps [6] in the more ambitious 2005 paper, thus no specialists in the area accept its widely publicized speculation.[7]

Lucio Russo has said that Plutarch, in his work *On the Face in the Moon*, was reporting some physical theories that we consider to be Newtonian and that these may have come originally from Hipparchus;[8] he goes on to say that Newton may have been influenced by them.[9] Both of these claims have been rejected by other scholars.[10]

A line in Plutarch's *Table Talk* states that Hipparchus counted 103049 compound propositions that can be formed from ten simple propositions; 103049 is the tenth Schröder–Hipparchus number and this line has led to speculation that Hipparchus knew about enumerative combinatorics, a field of mathematics that developed independently in modern mathematics.[11] [12]

Babylonian sources

Further information: Babylonian astronomy

Earlier Greek astronomers and mathematicians were influenced by Babylonian astronomy to some extent, for instance the period relations of the Metonic cycle and Saros cycle may have come from Babylonian sources (see "Babylonian astronomical diaries"). Hipparchus seems to have been the first to exploit Babylonian astronomical knowledge and techniques systematically.[13] Except for Timocharis and Aristillus, he was the first Greek known to divide the circle in 360 degrees of 60 arc minutes (Eratosthenes before him used a simpler sexagesimal system dividing a circle into 60 parts). He also used the Babylonian unit *pechus* ("cubit") of about 2° or 2.5°.

Hipparchus probably compiled a list of Babylonian astronomical observations; G. J. Toomer, a historian of astronomy, has suggested that Ptolemy's knowledge of eclipse records and other Babylonian observations in the *Almagest* came from a list made by Hipparchus. Hipparchus' use of Babylonian sources has always been known in a general way, because of Ptolemy's statements. However, Franz Xaver Kugler demonstrated that the synodic and anomalistic periods that Ptolemy attributes to Hipparchus had already been used in Babylonian ephemerides, specifically the collection of texts nowadays called "System B" (sometimes attributed to Kidinnu).[14]

Hipparchus's long draconitic lunar period (5458 months = 5923 lunar nodal periods) also appears a few times in Babylonian records. But the only such tablet explicitly dated is post-Hipparchus so the direction of transmission is not clear.

Geometry, trigonometry, and other mathematical techniques

Hipparchus was recognized as the first mathematician known to have possessed a trigonometric table, which he needed when computing the eccentricity of the orbits of the Moon and Sun. He tabulated values for the chord function, which gives the length of the chord for each angle. He did this for a circle with a circumference of 21600 and a radius (rounded) of 3438 units: this circle has a unit length of 1 arc minute along its perimeter. He tabulated the chords for angles with increments of 7.5°. In modern terms, the chord of an angle equals twice the sine of half of the angle, i.e.:

chord(A) = 2 sin(A/2).

He described the chord table in a work, now lost, called *Tōn en kuklō$_i$ eutheiōn* (*Of Lines Inside a Circle*) by Theon of Alexandria (4th century) in his commentary on the *Almagest* I.10; some claim his table may have survived in astronomical treatises in India, for instance the *Surya Siddhanta*. Trigonometry was a significant innovation, because it allowed Greek astronomers to solve any triangle, and made it possible to make quantitative astronomical models and predictions using their preferred geometric techniques.[15]

For his chord table Hipparchus must have used a better approximation for π than the one from Archimedes of between 3 + 1/7 and 3 + 10/71; perhaps he had the one later used by Ptolemy: 3;8:30 (sexagesimal) (*Almagest* VI.7); but it is not known if he computed an improved value himself.

But some scholars do not believe Arayabhatta's Sin table has anything to do with Hipparchus's chord table which does not exist today. Some scholars do not agree with this hypothesis that Hipparchus constructed a chord table. Bo. C Klintberg states " With mathematical reconstructions and philosophical arguments I show that Toomer's 1973 paper never contained any conclusive evidence for his claims that Hipparchus had a 3438'-based chord table, and that the Indians used that table to compute their sine tables. Recalculating Toomer's reconstructions with a 3600' radius -- i.e. the radius of the chord table in Ptolemy's Almagest, expressed in 'minutes' instead of 'degrees' -- generates Hipparchan-like ratios similar to those produced by a 3438' radius. It is therefore possible that the radius of Hipparchus's chord table was 3600', and that the Indians independently constructed their 3438'-based sine table." [16]

Hipparchus could construct his chord table using the Pythagorean theorem and a theorem known to Archimedes. He also might have developed and used the theorem in plane geometry called Ptolemy's theorem, because it was proved by Ptolemy in his *Almagest* (I.10) (later elaborated on by Carnot).

Hipparchus was the first to show that the stereographic projection is conformal, and that it transforms circles on the sphere that do not pass through the center of projection to circles on the plane. This was the basis for the astrolabe.

Besides geometry, Hipparchus also used arithmetic techniques developed by the Chaldeans. He was one of the first Greek mathematicians to do this, and in this way expanded the techniques available to astronomers and geographers.

There are several indications that Hipparchus knew spherical trigonometry, but the first surviving text of it is that of Menelaus of Alexandria in the 1st century, who on that basis is now commonly credited with its discovery. (Previous to the finding of the proofs of Menelaus a century ago, Ptolemy was credited with the invention of spherical trigonometry.) Ptolemy later used spherical trigonometry to compute things like the rising and setting points of the ecliptic, or to take account of the lunar parallax. Hipparchus may have used a globe for these tasks, reading values off coordinate grids drawn on it, or he may have made approximations from planar geometry, or perhaps used arithmetical approximations developed by the Chaldeans. He might have used spherical trigonometry.

Aubrey Diller has shown that the clima calculations which Strabo preserved from Hipparchus were performed by spherical trigonometry with the sole accurate obliquity known to have been used by ancient astronomers, 23°40'. All thirteen clima figures agree with Diller's proposal.[17] Further confirming his contention is the finding that the big errors in Hipparchus's longitude of Regulus and both longitudes of Spica agree to a few minutes in all three instances with a theory that he took the wrong sign for his correction for parallax when using eclipses for determining stars' positions.[18]

Lunar and solar theory

Motion of the Moon

Further information: Lunar theory and Orbit of the Moon

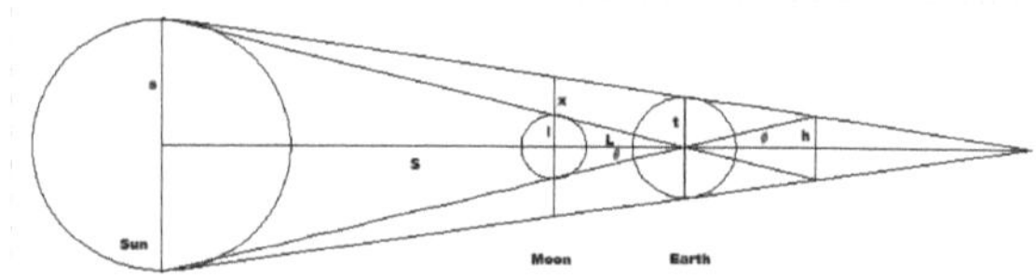

Geometric construction used by Hipparchus in his determination of the distances to the sun and moon.

Hipparchus also studied the motion of the Moon and confirmed the accurate values for two periods of its motion that Chaldean astronomers certainly possessed before him, whatever their ultimate origin. The traditional value (from Babylonian System B) for the mean synodic month is 29 days;31,50,8,20 (sexagesimal) = 29.5305941... days. Expressed as 29 days + 12 hours + 793/1080 hours this value has been used later in the Hebrew calendar (possibly from Babylonian sources). The Chaldeans also knew that 251 synodic months = 269 anomalistic months. Hipparchus used a multiple of this period by a factor of 17, because that interval is also an eclipse period. The Moon also is close to an integer number of years (4267 moons : 4573 anomalistic periods : 4630.53 nodal periods : 4611.98 lunar orbits : 344.996 years : 344.982 solar orbits : 126,007.003 days : 126,351.985 rotations). The 345-year eclipses reoccur with almost identical time of day, elevation, and celestial position.

Hipparchus could confirm his computations by comparing eclipses from his own time (presumably 27 January 141 BC and 26 November 139 BC according to [Toomer 1980]), with eclipses from Babylonian records 345 years earlier (*Almagest* IV.2; [A.Jones, 2001]). Already al-Biruni (*Qanun* VII.2.II) and Copernicus (*de revolutionibus* IV.4) noted that the period of 4,267 moons is actually about 5 minutes longer than the value for the eclipse period that Ptolemy attributes to Hipparchus. However, the timing methods of the Babylonians had an error of no less than 8 minutes.[19] Modern scholars agree that Hipparchus rounded the eclipse period to the nearest hour, and used it to confirm the validity of the traditional values, rather than try to derive an improved value from his own observations. From modern ephemerides [20] and taking account of the change in the length of the day (see ΔT) we estimate that the error in the assumed length of the synodic month was less than 0.2 seconds in the 4th century BC and less than 0.1 seconds in Hipparchus' time.

Orbit of the Moon

It had been known for a long time that the motion of the Moon is not uniform: its speed varies. This is called its *anomaly*, and it repeats with its own period; the anomalistic month. The Chaldeans took account of this arithmetically, and used a table giving the daily motion of the Moon according to the date within a long period. The Greeks however preferred to think in geometrical models of the sky. Apollonius of Perga had at the end of the 3rd century BC proposed two models for lunar and planetary motion:

1. In the first, the Moon would move uniformly along a circle, but the Earth would be eccentric, i.e., at some distance of the center of the circle. So the apparent angular speed of the Moon (and its distance) would vary.
2. The Moon itself would move uniformly (with some mean motion in anomaly) on a secondary circular orbit, called an *epicycle*, that itself would move uniformly (with some mean motion in longitude) over the main circular orbit around the Earth, called *deferent*; see deferent and epicycle.

Apollonius demonstrated that these two models were in fact mathematically equivalent. However, all this was theory and had not been put to practice. Hipparchus was the first astronomer we know attempted to determine the relative proportions and actual sizes of these orbits.

Hipparchus devised a geometrical method to find the parameters from three positions of the Moon, at particular phases of its anomaly. In fact, he did this separately for the eccentric and the epicycle model. Ptolemy describes the details in the *Almagest* IV.11. Hipparchus used two sets of three lunar eclipse observations, which he carefully selected to satisfy the requirements. The eccentric model he fitted to these eclipses from his Babylonian eclipse list: 22/23 December 383 BC, 18/19 June 382 BC, and 12/13 December 382 BC. The epicycle model he fitted to lunar eclipse observations made in Alexandria at 22 September 201 BC, 19 March 200 BC, and 11 September 200 BC.

- For the eccentric model, Hipparchus found for the ratio between the radius of the eccenter and the distance between the center of the eccenter and the center of the ecliptic (i.e., the observer on Earth): 3144 : 327+2/3 ;
- and for the epicycle model, the ratio between the radius of the deferent and the epicycle: 3122+1/2 : 247+1/2 .

The somewhat weird numbers are due to the cumbersome unit he used in his chord table according to one group of historians, who explain their reconstruction's inability to agree with these four numbers as partly due to some sloppy rounding and calculation errors by Hipparchus, for which Ptolemy criticised him (he himself made rounding errors too). A simpler alternate reconstruction [21] agrees with all four numbers. Anyway, Hipparchus found inconsistent results; he later used the ratio of the epicycle model (3122+1/2 : 247+1/2), which is too small (60 : 4;45 sexagesimal). Ptolemy established a ratio of 60 : 5+1/4.[22] (The maximum angular deviation producible by this geometry is the arcsin of 5 1/4 divided by 60, or about 5° 1', a figure that is sometimes therefore quoted as the equivalent of the Moon's equation of the center in the Hipparchan model.)

Apparent motion of the Sun

Before Hipparchus, Meton, Euctemon, and their pupils at Athens had made a solstice observation (i.e., timed the moment of the summer solstice) on 27 June 432 BC (proleptic Julian calendar). Aristarchus of Samos is said to have done so in 280 BC, and Hipparchus also had an observation by Archimedes. In 158 BC Hipparchus twice [23] computed from calendars an estimate for summer solstice. He observed the summer solstice in 146 BC and 135 BC both accurate [24] to a few hours, but observations of the moment of equinox were more precise, and he made twenty during his lifetime. Ptolemy gives an extensive discussion of Hipparchus' work on the length of the year in the *Almagest* III.1, and quotes many observations that Hipparchus made or used, spanning 162 BC to 128 BC. Analysis [25] of Hipparchus's seventeen equinox observations made at Rhodes shows that the mean error in declination is positive seven arc minutes, nearly agreeing with the sum of refraction by air and Swerdlow's parallax. The random noise is two arc minutes or more nearly one arcminute if rounding is taken into account which approximately agrees with the sharpness of the eye. Ptolemy quotes an equinox timing by Hipparchus (at 24 March 146 BC at dawn) that differs by 5 hours from the observation made on Alexandria's large public equatorial ring that same day (at 1 hour before noon): Hipparchus may have visited Alexandria but he did not make his equinox observations there; presumably he was on Rhodes (at nearly the same geographical longitude). He could have used the equatorial ring of his armillary sphere or another equatorial ring for these observations, but Hipparchus (and Ptolemy) knew that observations with these instruments are sensitive to a precise alignment with the equator, so if he were restricted to an armillary, it would make more sense to use its meridian ring as a transit instrument. The problem with an equatorial ring (if an observer is naive enough to trust it very near dawn or dusk) is that atmospheric refraction lifts the Sun significantly above the horizon: so for a northern hemisphere observer its apparent declination is too high, which changes the observed time when the Sun crosses the equator. (Worse, the refraction decreases as the Sun rises and increases as it sets, so it may appear to move in the wrong direction with respect to the equator in the course of the day – as Ptolemy mentions. Ptolemy and Hipparchus apparently did not realize that refraction is the cause.) However, such details have doubtful relation to the data of either man, since there is no textual, scientific, or statistical ground for believing that their equinoxes were taken on an equatorial ring, which is useless for solstices in any case. Not one of two centuries of mathematical investigations of their solar errors has claimed to have traced them to the effect of refraction on use of an equatorial ring. Ptolemy claims his solar observations were on a transit instrument set in the meridian.

At the end of his career, Hipparchus wrote a book called *Peri eniausíou megéthous* ("On the Length of the Year") about his results. The established value for the tropical year, introduced by Callippus in or before 330 BC was 365 + 1/4 days. Speculating a Babylonian origin for the Callippic year is hard to defend, since Babylon did not observe solstices thus the only extant System B yearlength was based on Greek solstices (see below). Hipparchus' equinox observations gave varying results, but he himself points out (quoted in *Almagest* III.1(H195)) that the observation errors by himself and his predecessors may have been as large as 1/4 day. He used old solstice observations, and determined a difference of about one day in about 300 years. So he set the length of the tropical year to 365 + 1/4 - 1/300 days (= 365.24666... days = 365 days 5 hours 55 min, which differs from the actual value (modern estimate) of 365.24219... days = 365 days 5 hours 48 min 45 s by only about 6 min).

Between the solstice observation of Meton and his own, there were 297 years spanning 108,478 days. D.Rawlins noted that this implies a tropical year of 365.24579... days = 365 days;14,44,51 (sexagesimal; = 365 days + 14/60 + $44/60^2$ + $51/60^3$) and that this exact yearlength has been found on one of the few Babylonian clay tablets which explicitly specifies the System B month. This is an indication that Hipparchus' work was known to Chaldeans.[26]

Another value for the year that is attributed to Hipparchus (by the astrologer Vettius Valens in the 1st century) is 365 + 1/4 + 1/288 days (= 365.25347... days = 365 days 6 hours 5 min), but this may be a corruption of another value attributed to a Babylonian source: 365 + 1/4 + 1/144 days (= 365.25694... days = 365 days 6 hours 10 min). It is not clear if this would be a value for the sidereal year (actual value at his time (modern estimate) about 365.2565 days), but the difference with Hipparchus' value for the tropical year is consistent with his rate of precession (see below).

Orbit of the Sun

Before Hipparchus, astronomers knew that the lengths of the seasons are not equal. Hipparchus made observations of equinox and solstice, and according to Ptolemy (*Almagest* III.4) determined that spring (from spring equinox to summer solstice) lasted 94½ days, and summer (from summer solstice to autumn equinox) 92½ days. This is inconsistent with a premise of the Sun moving around the Earth in a circle at uniform speed. Hipparchus' solution was to place the Earth not at the center of the Sun's motion, but at some distance from the center. This model described the apparent motion of the Sun fairly well. It is known today that the planets, including the Earth, move in approximate ellipses around the Sun, but this was not discovered until Johannes Kepler published his first two laws of planetary motion in 1609. The value for the eccentricity attributed to Hipparchus by Ptolemy is that the offset is 1/24 of the radius of the orbit (which is a little too large), and the direction of the apogee would be at longitude 65.5° from the vernal equinox. Hipparchus may also have used other sets of observations, which would lead to different values. One of his two eclipse trios' solar longitudes are consistent with his having initially adopted inaccurate lengths for spring and summer of 95¾ and 91¼ days.[27] His other triplet of solar positions is consistent with 94¼ and 92½ days,[28] an improvement on the results (94½ and 92½ days) attributed to Hipparchus by Ptolemy, which a few scholars still question the authorship of. Ptolemy made no change three centuries later, and expressed lengths for the autumn and winter seasons which were already implicit (as shown, e.g., by A. Aaboe).

Distance, parallax, size of the Moon and Sun

Hipparchus also undertook to find the distances and sizes of the Sun and the Moon. He published his results in a work of two books called *Perí megethōn kaí apostēmátōn* ("On Sizes and Distances") by Pappus in his commentary on the *Almagest* V.11; Theon of Smyrna (2nd century) mentions the work with the addition "of the Sun and Moon".

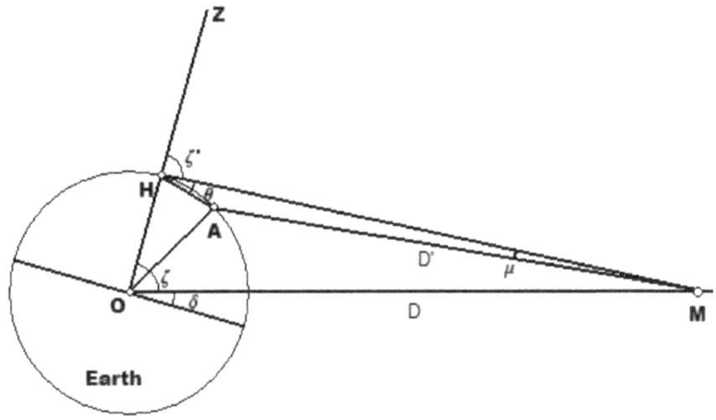

Diagram used in reconstructing one of Hipparchus' methods of determining the distance to the moon. This represents the earth-moon system during a partial solar eclipse at A (Alexandria) and a total solar eclipse at H (Hellespont).

Hipparchus measured the apparent diameters of the Sun and Moon with his *diopter*. Like others before and after him, he found that the Moon's size varies as it moves on its (eccentric) orbit, but he found no perceptible variation in the apparent diameter of the Sun. He found that at the **mean** distance of the Moon, the Sun and Moon had the same apparent diameter; at that distance, the Moon's diameter fits 650 times into the circle, i.e., the mean apparent diameters are 360/650 = 0°33'14".

Like others before and after him, he also noticed that the Moon has a noticeable parallax, i.e., that it appears displaced from its calculated position (compared to the Sun or stars), and the difference is greater when closer to the horizon. He knew that this is because in the then-current models the Moon circles the center of the Earth, but the observer is at the surface—the Moon, Earth and observer form a triangle with a sharp angle that changes all the time. From the size of this parallax, the distance of the Moon as measured in Earth radii can be determined. For the Sun however, there was no observable parallax (we now know that it is about 8.8", several times smaller than the resolution of the unaided eye).

In the first book, Hipparchus assumes that the parallax of the Sun is 0, as if it is at infinite distance. He then analyzed a solar eclipse, which Toomer (against the opinion of over a century of astronomers) presumes to be the eclipse of 14 March 190 BC. It was total in the region of the Hellespont (and in his birthplace, Nicaea); at the time Toomer proposes the Romans were preparing for war with Antiochus III in the area, and the eclipse is mentioned by Livy in his *Ab Urbe Condita* VIII.2. It was also observed in Alexandria, where the Sun was reported to be obscured 4/5ths by the Moon. Alexandria and Nicaea are on the same meridian. Alexandria is at about 31° North, and the region of the Hellespont about 40° North. (It has been contended that authors like Strabo and Ptolemy had fairly decent values for these geographical positions, so Hipparchus must have known them too. However, Strabo's Hipparchus dependent latitudes for this region are at least 1° too high, and Ptolemy appears to copy them, placing Byzantium 2° high in latitude.) Hipparchus could draw a triangle formed by the two places and the Moon, and from simple geometry was able to establish a distance of the Moon, expressed in Earth radii. Because the eclipse occurred in the morning, the Moon was not in the meridian, and it has been proposed that as a consequence the distance found by Hipparchus was a lower limit. In any case, according to Pappus, Hipparchus found that the least distance is 71 (from this eclipse), and the greatest 81 Earth radii.

In the second book, Hipparchus starts from the opposite extreme assumption: he assigns a (minimum) distance to the Sun of 490 Earth radii. This would correspond to a parallax of 7', which is apparently the greatest parallax that Hipparchus thought would not be noticed (for comparison: the typical resolution of the human eye is about 2'; Tycho Brahe made naked eye observation with an accuracy down to 1'). In this case, the shadow of the Earth is a cone rather than a cylinder as under the first assumption. Hipparchus observed (at lunar eclipses) that at the mean distance of the Moon, the diameter of the shadow cone is 2+½ lunar diameters. That apparent diameter is, as he had observed, 360/650 degrees. With these values and simple geometry, Hipparchus could determine the mean distance; because it was computed for a minimum distance of the Sun, it is the maximum mean distance possible for the Moon. With his value for the eccentricity of the orbit, he could compute the least and greatest distances of the Moon too. According to Pappus, he found a least distance of 62, a mean of 67+1/3, and consequently a greatest distance of 72+2/3 Earth radii. With this method, as the parallax of the Sun decreases (i.e., its distance increases), the minimum limit for the mean distance is 59 Earth radii – exactly the mean distance that Ptolemy later derived.

Hipparchus thus had the problematic result that his minimum distance (from book 1) was greater than his maximum mean distance (from book 2). He was intellectually honest about this discrepancy, and probably realized that especially the first method is very sensitive to the accuracy of the observations and parameters. (In fact, modern calculations show that the size of the 190 BC solar eclipse at Alexandria must have been closer to 9/10ths and not the reported 4/5ths, a fraction more closely matched by the degree of totality at Alexandria of eclipses occurring in 310 BC and 129 BC which were also nearly total in the Hellespont and are thought by many to be more likely possibilities for the eclipse Hipparchus used for his computations.)

Ptolemy later measured the lunar parallax directly (*Almagest* V.13), and used the second method of Hipparchus with lunar eclipses to compute the distance of the Sun (*Almagest* V.15). He criticizes Hipparchus for making contradictory assumptions, and obtaining conflicting results (*Almagest* V.11): but apparently he failed to understand Hipparchus' strategy to establish limits consistent with the observations, rather than a single value for the distance. His results were the best so far: the actual mean distance of the Moon is 60.3 Earth radii, within his limits from Hipparchus' second book.

Theon of Smyrna wrote that according to Hipparchus, the Sun is 1,880 times the size of the Earth, and the Earth twenty-seven times the size of the Moon; apparently this refers to volumes, not diameters. From the geometry of book 2 it follows that the Sun is at 2,550 Earth radii, and the mean distance of the Moon is 60½ radii. Similarly, Cleomedes quotes Hipparchus for the sizes of the Sun and Earth as 1050:1; this leads to a mean lunar distance of 61 radii. Apparently Hipparchus later refined his computations, and derived accurate single values that he could use for predictions of solar eclipses.

See [Toomer 1974] for a more detailed discussion.

Eclipses

Pliny (*Naturalis Historia* II.X) tells us that Hipparchus demonstrated that lunar eclipses can occur five months apart, and solar eclipses seven months (instead of the usual six months); and the Sun can be hidden twice in thirty days, but as seen by different nations. Ptolemy discussed this a century later at length in *Almagest* VI.6. The geometry, and the limits of the positions of Sun and Moon when a solar or lunar eclipse is possible, are explained in *Almagest* VI.5. Hipparchus apparently made similar calculations. The result that two solar eclipses can occur one month apart is important, because this can not be based on observations: one is visible on the northern and the other on the southern hemisphere – as Pliny indicates – and the latter was inaccessible to the Greek.

Prediction of a solar eclipse, i.e., exactly when and where it will be visible, requires a solid lunar theory and proper treatment of the lunar parallax. Hipparchus must have been the first to be able to do this. A rigorous treatment requires spherical trigonometry, thus those who remain certain that Hipparchus lacked it must speculate that he may have made do with planar approximations. He may have discussed these things in *Perí tēs katá plátos mēniaías tēs selēnēs kinēseōs* ("On the monthly motion of the Moon in latitude"), a work mentioned in the *Suda*.

Pliny also remarks that "he also discovered for what exact reason, although the shadow causing the eclipse must from sunrise onward be below the earth, it happened once in the past that the moon was eclipsed in the west while both luminaries were visible above the earth" (translation H. Rackham (1938), Loeb Classical Library 330 p. 207). Toomer (1980) argued that this must refer to the large total lunar eclipse of 26 November 139 BC, when over a clean sea horizon as seen from Rhodes, the Moon was eclipsed in the northwest just after the Sun rose in the southeast.

This would be the second eclipse of the 345-year interval that Hipparchus used to verify the traditional Babylonian periods: this puts a late date to the development of Hipparchus' lunar theory. We do not know what "exact reason" Hipparchus found for seeing the Moon eclipsed while apparently it was not in exact opposition to the Sun. Parallax lowers the altitude of the luminaries; refraction raises them, and from a high point of view the horizon is lowered.

Astronomical instruments and astrometry

Hipparchus and his predecessors used various instruments for astronomical calculations and observations, such as the gnomon, the astrolabe, and the armillary sphere.

Hipparchus is credited with the invention or improvement of several astronomical instruments, which were used for a long time for naked-eye observations. According to Synesius of Ptolemais (4th century) he made the first *astrolabion*: this may have been an armillary sphere (which Ptolemy however says he constructed, in *Almagest* V.1); or the predecessor of the planar instrument called astrolabe (also mentioned by Theon of Alexandria). With an astrolabe Hipparchus was the first to be able to measure the geographical latitude and time by observing stars. Previously this was done at daytime by measuring the shadow cast by a *gnomon*, or with the portable instrument known as a *scaphe*.

Ptolemy mentions (*Almagest* V.14) that he used a similar instrument as Hipparchus, called *dioptra*, to measure the apparent diameter of the Sun and Moon. Pappus of Alexandria described it (in his commentary on the *Almagest* of that chapter), as did Proclus (*Hypotyposis* IV). It was a 4-foot rod with a scale, a sighting hole at one end, and a wedge that could be moved along the rod to exactly obscure the disk of Sun or Moon.

Hipparchus also observed solar equinoxes, which may be done with an equatorial ring: its shadow falls on itself when the Sun is on the equator (i.e., in one of the equinoctial points on the ecliptic), but the shadow falls above or below the opposite side of the ring when the Sun is south or north of the equator. Ptolemy quotes (in *Almagest* III.1 (H195)) a description by Hipparchus of an equatorial ring in Alexandria; a little further he describes two such instruments present in Alexandria in his own time.

Hipparchus applied his knowledge of spherical angles to the problem of denoting locations on the Earth's surface. Before him a grid system had been used by Dicaearchus of Messana, but Hipparchus was the first to apply mathematical rigor to the determination of the latitude and longitude of places on the Earth. Hipparchus wrote a critique in three books on the work of the geographer Eratosthenes of Cyrene (3rd century BC), called *Pròs tèn 'Eratosthénous geografían* ("Against the Geography of Eratosthenes"). It is known to us from Strabo of Amaseia, who in his turn criticised Hipparchus in his own *Geografia*. Hipparchus apparently made many detailed corrections to the locations and distances mentioned by Eratosthenes. It seems he did not introduce many improvements in methods, but he did propose a means to determine the geographical longitudes of different cities at lunar eclipses (Strabo *Geografia* 1 January 2012). A lunar eclipse is visible simultaneously on half of the Earth, and the difference in longitude between places can be computed from the difference in local time when the eclipse is observed. His approach would give accurate results if it were correctly carried out but the limitations of timekeeping accuracy in his era made this method impractical.

Star catalog

Late in his career (possibly about 135 BC) Hipparchus compiled his star catalog, the original of which does not survive. He also constructed a celestial globe depicting the constellations, based on his observations. His interest in the fixed stars may have been inspired by the observation of a supernova (according to Pliny), or by his discovery of precession, according to Ptolemy, who says that Hipparchus could not reconcile his data with earlier observations made by Timocharis and Aristillus. For more information see Discovery of precession.

Previously, Eudoxus of Cnidus in the 4th century BC had described the stars and constellations in two books called *Phaenomena* and *Entropon*. Aratus wrote a poem called *Phaenomena* or *Arateia* based on Eudoxus' work.

Hipparchus wrote a commentary on the *Arateia* – his only preserved work – which contains many stellar positions and times for rising, culmination, and setting of the constellations, and these are likely to have been based on his own measurements.

Hipparchus made his measurements with an armillary sphere, and obtained the positions of at least 850 stars. It is disputed which coordinate system(s) he used. Ptolemy's catalog in the *Almagest*, which is derived from Hipparchus' catalog, is given in ecliptic coordinates. However Delambre in his *Histoire de l'Astronomie Ancienne* (1817) concluded that Hipparchus knew and used the equatorial coordinate system, a conclusion challenged by Otto Neugebauer in his *A History of Ancient Mathematical Astronomy* (1975). Hipparchus seems to have used a mix of ecliptic coordinates and equatorial coordinates: in his commentary on Eudoxos he provides stars' polar distance (equivalent to the declination in the equatorial system), right ascension (equatorial), longitude (ecliptical), polar longitude (hybrid), but not celestial latitude.

As with most of his work, Hipparchus' star catalog was adopted and perhaps expanded by Ptolemy. Delambre, in 1817, cast doubt on Ptolemy's work. It was disputed whether the star catalog in the *Almagest* is due to Hipparchus, but 1976–2002 statistical and spatial analyses (by R. R. Newton, Dennis Rawlins, Gerd Grasshoff,[29] Keith Pickering[30] and Dennis Duke[31]) have shown conclusively that the *Almagest* star catalog is almost entirely Hipparchan. Ptolemy has even (since Brahe, 1598) been accused by astronomers of fraud for stating (*Syntaxis*, book 7, chapter 4) that he observed all 1025 stars: for almost every star he used Hipparchus' data and precessed it to his own epoch 2⅔ centuries later by adding 2°40' to the longitude, using an erroneously small precession constant of 1° per century.

In any case the work started by Hipparchus has had a lasting heritage, and was much later updated by Al Sufi (964) and Copernicus (1543). Ulugh Beg reobserved all the Hipparchus stars he could see from Samarkand in 1437 to about the same accuracy as Hipparchus's. The catalog was superseded only in the late 16th century by Brahe and Wilhelm IV of Kassel via superior ruled instruments and spherical trigonometry, which improved accuracy by an order of magnitude even before the invention of the telescope. Hipparchus is considered the greatest observational astronomer from classical antiquity until Brahe.[32]

Stellar magnitude

Hipparchus ranked stars in six magnitude classes according to their brightness: he assigned the value of one to the twenty brightest stars, to fainter ones a value of two, and so forth to the stars with a class of six, which can be barely seen with the naked eye. That system, improved, is still used today.

Precession of the equinoxes (146–127 BC)

See also Precession (astronomy)

Hipparchus is known for being almost universally recognized as discoverer of the precession of the equinoxes in 127 BC.[33] His two books on precession, *On the Displacement of the Solsticial and Equinoctial Points* and *On the Length of the Year*, are both mentioned in the *Almagest* of Claudius Ptolemy. According to Ptolemy, Hipparchus measured the longitude of Spica and Regulus and other bright stars. Comparing his measurements with data from his predecessors, Timocharis and Aristillus, he concluded that Spica had moved 2° relative to the autumnal equinox. He also compared the lengths of the tropical year (the time it takes the Sun to return to an equinox) and the sidereal year (the time it takes the Sun to return to a fixed star), and found a slight discrepancy. Hipparchus concluded that the equinoxes were moving ("precessing") through the zodiac, and that the rate of precession was not less than 1° in a century.

Geography

Hipparchus' treatise "Against the Geography of Eratosthenes" in three books is not preserved.[34] Most of our knowledge of it comes from Strabo. Hipparchus thoroughly and often unfairly criticized Eratosthenes mainly for internal contradictions and inaccuracy in determining positions of geographical localities. Hipparchus insists that a geographic map must be based only on astronomical measurements of latitudes and longitudes and triangulation for finding unknown distances. In geographic theory and methods Hipparchus introduced three main innovations.[35] He was the first to use the grade grid, to determine geographic latitude from star observations, and not only from the sun's altitude, a method known long before him, and to suggest that geographic longitude could be determined by means of simultaneous observations of lunar eclipses in distant places. In the practical part of his work, the so-called "table of climata", Hipparchus listed latitudes for several tens of localities. In particular, he improved Eratosthenes' values for the latitudes of Athens, Sicily, and southern extremity of India.[36] In calculating latitudes of climata (latitudes correlated with the length of the longest solstitial day), Hipparchus used an unexpectedly accurate value for the obliquity of the ecliptic, 23°40′ (the actual value in the second half of the 2nd cent. B.C. was approximately 23°43′), whereas all other ancient authors knew only a roughly rounded value 24°, and even Ptolemy used a less accurate value, 23°51′.[37] Hipparchus opposed the view generally accepted in the Hellenistic period that the Atlantic and Indian Oceans and the Caspian Sea are parts of a single ocean. At the same time he extends the limits of the oikoumene, i.e. the inhabited part of the land, up to the equator and the Arctic Circle.[38] These Hipparchus' ideas found their reflection in the Geography of Ptolemy. In essence, Ptolemy's work is an extended attempt to realize Hipparchus' vision of what geography ought to be.

Named after Hipparchus

The rather cumbersome formal name for the ESA's Hipparcos Space Astrometry Mission was High Precision Parallax Collecting Satellite; it was deliberately named in this way to give an acronym, HiPParCoS, that echoed and commemorated the name of Hipparchus. The lunar crater Hipparchus and the asteroid 4000 Hipparchus are more directly named after him.

Monument

The Astronomer's Monument at the Griffith Observatory in Los Angeles, California, USA features a relief of Hipparchus as one of six of the greatest astronomers of all time and the only one from Antiquity.

Notes

[1] C. M. Linton (2004). *From Eudoxus to Einstein: a history of mathematical astronomy*. Cambridge University Press. p. 52. ISBN 0-521-82750-7.

[2] G J Toomer's chapter "Ptolemy and his Greek Predecessors" in "Astronomy Before the Telescope", British Museum Press, 1996, p 81

[3] *Astronomies and cultures in early medieval Europe, Stephen C. McCluskey, Cambridge University Press, 2000, p. 22* (http://books.google.com/books?id=PsLArVtOsPYC&printsec=frontcover&dq="astronomies+and+Cultures+in+early+medieval+europe"&source=bl&ots=R9abnu2Aah&sig=TRhYkhhjP0rgI8XzJceVPcKL2Uc&hl=en&ei=jMXtTO7KDIT48Aaxj8moDA&sa=X&oi=book_result&ct=result&resnum=2&ved=0CD0Q6AEwAQ#v=onepage&q=ptolemy&f=false). Books.google.com. 2000-11-20. . Retrieved 2012-08-06.

[4] For general information on Hipparchus see the following biographical articles: G. J. Toomer, "Hipparchus" (1978); and A. Jones, "Hipparchus."

[5] Modern edition: Karl Manitius (*In Arati et Eudoxi Phaenomena*, Leipzig, 1894).

[6] D.Rawlins (http://www.dioi.org/ggg.htm), "Farnese Atlas Celestial Globe, Proposed Astronomical Origins", 2005.

[7] B. E. Schaefer (http://www.phys.lsu.edu/farnese/), "Epoch of the Constellations on the Farnese Atlas and their Origin in Hipparchus's Lost Catalog", *Journal for the History of Astronomy*, May, 2005 versus Dennis Duke (http://www.people.scs.fsu.edu/~dduke/farnese) *Journal for the History of Astronomy*, February, 2006.

[8] Lucio Russo, *The Forgotten Revolution: How Science Was Born in 300 BC and Why It Had To Be Reborn*, (Berlin: Springer, 2004). ISBN 3-540-20396-6, pp. 286–293.

[9] Lucio Russo, *The Forgotten Revolution: How Science Was Born in 300 BC and Why It Had To Be Reborn*, (Berlin: Springer, 2004). ISBN 3-540-20396-6, pp. 365–379.

[10] Mott Greene, "The birth of modern science?" Review of *The Forgotten Revolution*, *Nature* **430** (5 August 2004): 614.

[11] Stanley, Richard P. (1997), "Hipparchus, Plutarch, Schröder, and Hough" (http://www-math.mit.edu/~rstan/papers/hip.pdf), *The American Mathematical Monthly* **104** (4): 344–350, doi:10.2307/2974582, MR1450667,

[12] Acerbi, F. (2003), "On the shoulders of Hipparchus: A reappraisal of ancient Greek combinatorics" (http://stl.recherche.univ-lille3.fr/sitespersonnels/acerbi/acerbipub5.pdf), *Archive for History of Exact Sciences* **57**: 465–502,

[13] For more information see G. J. Toomer, "Hipparchus and Babylonian astronomy."

[14] Franz Xaver Kugler, *Die Babylonische Mondrechnung* ("The Babylonian lunar computation"), Freiburg im Breisgau, 1900.

[15] Toomer, "The Chord Table of Hipparchus" (1973).

[16] Reference: Hipparchus's 3600'-Based Chord Table and Its Place in the History of Ancient Greek and Indian Trigonometry, Bo C. Klintberg, Indian Journal of History of Science 40 (2):169-203 (2005)

[17] Dennis Rawlins, "Aubrey Diller Legacies" (http://www.dioi.org/vols/w50.pdf), *DIO* 5 (2009); Shcheglov D.A. (2002-2007): "Hipparchus' Table of Climata and Ptolemy's Geography" (http://nw.academia.edu/DmitryShcheglov/Papers/142876/Hipparchus_Table_of_Climata_and_Ptolemys_Geography), Orbis Terrarum 9 (2003–2007), 177–180.

[18] Dennis Rawlins, "Ancient spherical trig" (http://www.dioi.org/vols/wg0.pdf), *DIO* 16 (2009).

[19] Stephenson & Fatoohi 1993; Steele *et al.* 1997

[20] Chapront *et al.* 2002

[21] Summarized in Hugh Thurston (2002): *Isis* **93**, 58–69.

[22] Toomer, 1967

[23] http://www.dioi.org/cot.htm#nthq

[24] http://www.dioi.org/cot.htm#htqp

[25] http://www.dioi.org/cot.htm#qxhr

[26] *DIO* (http://www.dioi.org/vols/w11.pdf), volume 1, number 1, pages 49–66; A.Jones, 2001; Thurston, *op. cit.*, page 62

[27] Thurston, *op. cit.*, page 67, note 16. R.Newton proposed that Hipparchus made an error of a degree in one of the trios' eclipses. D.Rawlins's theory (Thurston *op. cit.*) that Hipparchus analysed the two trios in pairs not threesomes provides a possible explanation for the one degree slip. It was a fudge (http://www.dioi.org/cot.htm#qtpt) necessitated by inadequacies of analysing by pairs instead of using the better method Ptolemy applies at *Almagest* Book 4 Parts 6 and 11.

[28] *Ibid*, note 14; Jones 2001

[29] Gerd Grasshoff: The history of Ptolemy's star catalogue, Springer, New York, 1990, ISBN 3-540-97181-5 (Analyse des im "Almagest" überlieferten Sternenkatalogs)

[30] "Keith Pickering" (http://www.dioi.org/vols/wc0.pdf) (PDF). . Retrieved 2012-08-06.

[31] "The Measurement Method of the Almagest Stars" (http://people.sc.fsu.edu/~dduke/model2.pdf), by Dennis Duke (http://people.scs.fsu.edu/~dduke), DIO: the International Journal of Scientific History,12 (2002).

[32] Benson Bobrick, "The Fated Sky", Simon & Schuster, 2005, p 151

[33] Giorgio de Santillana & Hertha von Dechend, "Hamlet's Mill", David R Godine, Boston, publisher, 1977, p 66

[34] Editions of fragments: Berger H. *Die geographischen Fragmente des Hipparch* (http://books.google.ru/books?id=1qNDAAAAcAAJ&printsec=frontcover&dq=berger+hugo+hipparchus&source=bl&ots=4yHbGolIRN&sig=gOAsU8DfieuAUEJKiHO4bW198nA&hl=ru&sa=X&ei=telmUKShJ66N4gSTo4GoAQ&ved=0CDQQ6AEwAQ#v=onepage&q=berger hugo hipparchus&f=false). Leipzig: B. G. Teubner, 1869.; Dicks D.R. *The Geographical Fragments of Hipparchus*. London: Athlon Press, 1960.

[35] On Hipparchus' geography see: Berger H. *Die geographischen Fragmente des Hipparch* (http://books.google.ru/books?id=1qNDAAAAcAAJ&printsec=frontcover&dq=berger+hugo+hipparchus&source=bl&ots=4yHbGolIRN&sig=gOAsU8DfieuAUEJKiHO4bW198nA&hl=ru&sa=X&ei=telmUKShJ66N4gSTo4GoAQ&ved=0CDQQ6AEwAQ#v=onepage&q=berger hugo hipparchus&f=false). Leipzig: B. G. Teubner, 1869.; Dicks D.R. *The Geographical Fragments of Hipparchus. London: Athlon Press, 1960; Neugebauer O.* A History of Ancient Mathematical Astronomy. *Pt. 1-3. Berlin, Heidelberg, New York: Springer Verlag, 1975: 332–338;* Shcheglov D.A. *Hipparchus' "Table of Climata and Ptolemy's Geography"* (http://nw.academia.edu/DmitryShcheglov/Papers/142876/Hipparchus_Table_of_Climata_and_Ptolemys_Geography). Orbis Terrarum *9. 2003–2007: 159–192.*

[36] Shcheglov D.A. "Hipparchus on the Latitude of Southern India" (http://nw.academia.edu/DmitryShcheglov/Papers/110297/Hipparchus_on_the_Latitude_of_Southern_India). *Greek, Roman, and Byzantine Studies* 45. 2005: 359–380; idem. " Eratosthenes' Parallel of Rhodes and the History of the System of Climata (http://www.academia.edu/191065/Eratosthenes_Parallel_of_Rhodes_and_the_History_of_the_System_of_Climata)". *Klio* 88. 2006: 351–359.; idem. "Hipparchus' Table of Climata and Ptolemy's Geography" (http://nw.academia.edu/DmitryShcheglov/Papers/142876/Hipparchus_Table_of_Climata_and_Ptolemys_Geography). *Orbis Terrarum* 9. 2003–2007: 159–192.

[37] Diller A. (1934). "Geographical Latitudes in Eratosthenes, Hipparchus and Posidonius". *Klio* 27.3: 258–269; cf. Shcheglov D.A. "Hipparchus' Table of Climata and Ptolemy's Geography" (http://nw.academia.edu/DmitryShcheglov/Papers/142876/Hipparchus_Table_of_Climata_and_Ptolemys_Geography), 177–180.

[38] Shcheglov D.A. "Ptolemy's Latitude of Thule and the Map Projection in the Pre-Ptolemaic Geography (http://nw.academia.edu/DmitryShcheglov/Papers/141379/Ptolemys_Latitude_of_Thule_and_the_Map_Projection_in_the_Pre-Ptolemaic_Geography). *Antike Naturwissenschaft und ihre Rezeption (AKAN)* 17. 2007: 132-139.

Editions and translations

- Berger H. *Die geographischen Fragmente des Hipparch.* Leipzig: B. G. Teubner, 1869 (http://books.google.ru/books?id=1qNDAAAAcAAJ&printsec=frontcover&dq=berger+hugo+hipparchus&source=bl&ots=4yHbGolIRN&sig=gOAsU8DfieuAUEJKiHO4bW198nA&hl=ru&sa=X&ei=telmUKShJ66N4gSTo4GoAQ&ved=0CDQQ6AEwAQ#v=onepage&q=berger hugo hipparchus&f=false).
- Dicks D.R. *The Geographical Fragments of Hipparchus.* Edited with an Introduction and Commentary. London: Athlon Press, 1960. Pp. xi + 215.
- Manitius K. *In Arati et Eudoxi Phaenomena commentariorum libri tres.* Leipzig: B. G. Teubner, 1894. 376 S.

Bibliography

- Acerbi F. (2003). "On the shoulders of Hipparchus: A reappraisal of ancient Greek combinatorics" (http://stl.recherche.univ-lille3.fr/sitespersonnels/acerbi/acerbipub5.pdf). *Archive for History of Exact Sciences* **57**: 465–502.
- Bianchetti S. (2001). "Dall'astronomia alla cartografia: Ipparco di Nicea". *ΠΟΙΚΙΛΜΑ. Studi in onore di Michelle R. Cataudella in occasione del 60° compleanno.* La Spezia: Agorà Edizioni: 145–156.
- Bowen A.C., Goldstein B.R. (1991). "Hipparchus' Treatment of Early Greek Astronomy: The Case of Eudoxus and the Length of Daytime Author(s)". *Proceedings of the American Philosophical Society* **135(2)**: 233-254.
- Chapront J., Touze M. Chapront, Francou G. (2002): "A new determination of lunar orbital parameters, precession constant, and tidal acceleration from LLR measurements" (http://adsabs.harvard.edu/abs/2002A&A...387..700C). *Astronomy and Astrophysics* **387**: 700–709.
- Dicks D.R. (1960). *The Geographical Fragments of Hipparchus.* London: Athlon Press. Pp. xi, 215.
- Diller A. (1934). "Geographical Latitudes in Eratosthenes, Hipparchus and Posidonius". *Klio* 27(3): 258–269.
- Duke D.W. (2002). "Associations between the ancient star catalogs". *Archive for the History of Exact Sciences* 56(5):435–450. (Author's draft here.) (http://people.sc.fsu.edu/~dduke/assocs.pdf)
- Honigmann E. (1929). *Die sieben Klimata und die πολεις επισημοι.* Eine Untersuchung zur Geschichte der Geographie und Astrologie in Altertum und Mittelalter. Heidelberg: Carl Winter's Universitätsbuchhandlung. 247 S.
- Jones A. (2001). "Hipparchus." In *Encyclopedia of Astronomy and Astrophysics.* Nature Publishing Group.
- Moore P. (1994). *Atlas of the Universe,* Octopus Publishing Group LTD (Slovene translation and completion by Tomaž Zwitter and Savina Zwitter (1999): *Atlas vesolja*): 225.
- Nadal R., Brunet J.P. (1984). "Le "Commentaire" d'Hipparque. I. La sphère mobile. *Archive for History of Exact Sciences* **29**: 201–236.
- Neugebauer O. (1975). *A History of Ancient Mathematical Astronomy.* Vol. 1-3. Berlin, Heidelberg, New York: Springer Verlag.
- Newton R.R. (1977). *The Crime of Claudius Ptolemy.* Baltimore: Johns Hopkins University Press.
- Rawlins D. (1982). An Investigation of the Ancient Star Catalog. *Proceedings of the Astronomical Society of the Pacific* 94, 359–373. Has been updated several times: *DIO* (http://www.dioi.org/vols/w82.pdf), volume 8, number 1 (1998), page 2, note 3, and *DIO* (http://www.dioi.org/vols/wa0.pdf), volume 10 (2000), page 79, note 177.
- Russo L. (1994). "The astronomy of Hipparchus and his time: A study based on pre-ptolemaic sources". *Vistas in Astronomy* **38.2**: 207—248
- Schaefer B.E. (2005). "The Epoch of the Constellations on the Farnese Atlas and their Origin in Hipparchus's Lost Catalogue" (http://articles.adsabs.harvard.edu/full/2005JHA....36..167S). *Journal for the History of Astronomy* **36.2**: 167-196.
- Shcheglov D.A. (2005). "Hipparchus on the Latitude of Southern India" (http://nw.academia.edu/DmitryShcheglov/Papers/110297/Hipparchus_on_the_Latitude_of_Southern_India). *Greek, Roman, and*

Byzantine Studies **45**: 359–380.

- Shcheglov D.A. (2006). " Eratosthenes' Parallel of Rhodes and the History of the System of Climata (http://www.academia.edu/191065/Eratosthenes_Parallel_of_Rhodes_and_the_History_of_the_System_of_Climata)". *Klio* **88**: 351–359.
- Shcheglov D.A. (2007). "Ptolemy's Latitude of Thule and the Map Projection in the Pre-Ptolemaic Geography" (http://nw.academia.edu/DmitryShcheglov/Papers/141379/Ptolemys_Latitude_of_Thule_and_the_Map_Projection_in_the_Pre-Ptolemaic_Geography). *Antike Naturwissenschaft und ihre Rezeption* (AKAN) **17**: 121–151.
- Shcheglov D.A. (2003-2007). "Hipparchus' Table of Climata and Ptolemy's Geography" (http://nw.academia.edu/DmitryShcheglov/Papers/142876/Hipparchus_Table_of_Climata_and_Ptolemys_Geography). *Orbis Terrarum* **9**: 159–192.
- Sidoli N. (2004). "Hipparchus and the Ancient Metrical Methods on the Sphere". *Journal for the History of Astronomy* 35: 71-84.
- Steele J.M., Stephenson F.R., Morrison L.V. (1997). "The accuracy of eclipse times measured by the Babylonians" (http://articles.adsabs.harvard.edu/full/1997JHA....28..337S). *Journal for the History of Astronomy* **28**, 337..345
- Stephenson F.R., Fatoohi L.J. (1993). "Lunar Eclipse Times Recorded in Babylonian History" (http://articles.adsabs.harvard.edu/full/1993JHA....24..255S). *Journal for the History of Astronomy* **24**: 255..267
- Swerdlow N.M. (1969). "Hipparchus on the distance of the sun." *Centaurus* **14**: 287–305.
- Toomer G.J. (1967). "The Size of the Lunar Epicycle According to Hipparchus." *Centaurus* **12**: 145–150.
- Toomer G.J. (1973). "The Chord Table of Hipparchus and the Early History of Greek Trigonometry." *Centaurus* **18**: 6–28.
- Toomer G.J. (1974). "Hipparchus on the Distances of the Sun and Moon." *Archives for the History of the Exact Sciences* **14**: 126–142.
- Toomer G.J. (1978). "Hipparchus." In *Dictionary of Scientific Biography* **15**: 207–224.
- Toomer G.J. (1980). "Hipparchus' Empirical Basis for his Lunar Mean Motions," *Centaurus* **24**: 97–109.
- Toomer G.J. (1988). "Hipparchus and Babylonian Astronomy." In *A Scientific Humanist: Studies in Memory of Abraham Sachs*, ed. Erle Leichty, Maria deJ. Ellis, and Pamel Gerardi. Philadelphia: Occasional Publications of the Samuel Noah Kramer Fund, 9.
- Wolff M. (1989). "Hipparchus and the Stoic Theory of Motion". In *Matter and Metaphysics*. Ed. J. Barnes & M. Mignucci. Napoli: Bibliopolis: 346—419.

Further reading

- Dreyer, John L.E (1953). *A History of Astronomy from Thales to Kepler*. New York: Dover Publications.
- Heath, Thomas (1921). *A History of Greek Mathematics* (http://books.google.com/books?id=LOA5AAAAMAAJ). Oxford: Clarendon Press.
- Lloyd, G.E.R. (1973). *Greek science after Aristotle*. New York: Norton. ISBN 0-393-04371-1.
- Neugebauer, Otto (1956). "Notes on Hipparchus". In Weinberg, Saul S. *The Aegean and the Near East: Studies Presented to Hetty Goldman*. Locust Valley, NY: J.J. Augustin.
- Ptolemy (1984). *Ptolemy's Almagest*. G.J. Toomer, trans. New York: Springer-Verlag. ISBN 0-387-91220-7.
- Thomson, J.Oliver (1948). *History of Ancient Geography*. Cambridge: Cambridge University Press.

External links

General

- O'Connor, John J.; Robertson, Edmund F., "Hipparchus" (http://www-history.mcs.st-andrews.ac.uk/Biographies/Hipparchus.html), *MacTutor History of Mathematics archive*, University of St Andrews.
- Biographical page at the (http://www.hps.cam.ac.uk/starry/hipparchus.html) University of Cambridge
- University of Cambridge's Page about Hipparchus' sole surviving work (http://www.hps.cam.ac.uk/starry/hipppoem.html)
- Biographical page at the (http://zebu.uoregon.edu/~js/glossary/hipparchus.html) University of Oregon
- Biography of Hipparchus on Fermat's Last Theorem Blog (http://fermatslasttheorem.blogspot.com/2006/04/hipparchus-of-rhodes.html)
- Pastore, Giovanni - *THE RECOVERED ARCHIMEDES' PLANETARIUM*, Rome 2010, ISBN 978-88-904715-2-0 (http://www.giovannipastore.it/ARCHIMEDE.htm)
- The Antikythera Calculator (Italian and English versions) (http://www.giovannipastore.it/ANTIKYTHERA.htm)
- Os Eclipses, AsterDomus website, portuguese (http://www.asterdomus.com.br/Artigo_os_eclipses.htm)
- Ancient Astronomy, Integers, Great Ratios, and Aristarchus (http://jqjacobs.net/astro/aristarchus.html)

Precession

- David Ulansey about Hipparchus's understanding of the precession (http://www.well.com/user/davidu/appendix4.html)

Celestial bodies

- M44 Praesepe at SEDS (http://www.seds.org/messier/m/m044.html), University of Arizona

Star catalog

- A brief view by Carmen Rush on Hipparchus' stellar catalog (http://ottawa.rasc.ca/astronomy/astro_facts/hipparcus.html)

Hero_of_Alexandria

Heron

Born	c. 10 AD
Residence	Alexandria, Roman Egypt
Fields	Engineering mathematics
Known for	*aeolipile*

Hero (or **Heron**) **of Alexandria** (Greek: **Ἥρων ὁ Ἀλεξανδρεύς**) (c. 10–70 AD) was an ancient Greek mathematician and engineer[1] [2] [3] who was active in his native city of Alexandria, Roman Egypt. He is considered the greatest experimenter of antiquity[1] and his work is representative of the Hellenistic scientific tradition.[4]

Hero published a well recognized description of a steam-powered device called an *aeolipile* (hence sometimes called a "Hero engine"). Among his most famous inventions was a windwheel, constituting the earliest instance of wind harnessing on land.[5] [6] He is said to have been a follower of the Atomists. Some of his ideas were derived from the works of Ctesibius.

Much of Hero's original writings and designs have been lost, but some of his works were preserved in Arab manuscripts.

Career

A number of references mention dates around 150 BC, but these are inconsistent with the dates of his publications and inventions. This may be due to a misinterpretation of the phrase "first century" or because Hero was a common name.

It is almost certain that Hero taught at the Musaeum which included the famous Library of Alexandria, because most of his writings appear as lecture notes for courses in mathematics, mechanics, physics and pneumatics. Although the field was not formalized until the 20th century, it is thought that the work of Hero, his automated devices in particular, represents some of the first formal research into cybernetics.[7]

Inventions and achievements

Heron's Aeolipile

- Hero described construction of the *aeolipile* (a version of which is known as *Hero's engine*) which was a rocket-like reaction engine and the first-recorded steam engine (although Vitruvius mentioned the aeolipile in De Architectura some 100 years earlier than Hero). It was created almost two millennia before the industrial revolution. Another engine used air from a closed chamber heated by an altar fire to displace water from a sealed vessel; the water was collected and its weight, pulling on a rope, opened temple doors.[8] Some historians have conflated the two inventions to assert that the aeolipile was capable of useful work.[9]
- The first vending machine was also one of his constructions, when a coin was introduced via a slot on the top of the machine, a set amount of holy water was dispensed. This was included in his list of inventions in his book, "Mechanics and Optics". When the coin was deposited, it fell upon a pan attached to a lever. The lever opened up a valve which let some water flow out. The pan continued to tilt with the weight of the coin until it fell off, at which point a counter-weight would snap the lever back up and turn off the valve.[10]
- A windwheel operating an organ, marking the first instance of wind powering a machine in history.[5] [6]
- Hero also invented many mechanisms for the Greek theater, including an entirely mechanical play almost ten minutes in length, powered by a binary-like system of ropes, knots, and simple machines operated by a rotating cylindrical cogwheel. The sound of thunder was produced by the mechanically-timed dropping of metal balls onto a hidden drum.
- The force pump was widely used in the Roman world, and one application was in a fire-engine.
- A syringe-like device was described by Heron to control the delivery of air or liquids.[11]
- In optics, Hero formulated the Principle of the Shortest Path of Light: If a ray of light propagates from point A to point B within the same medium, the path-length followed is the shortest possible. It was nearly 1000 years later that Alhacen expanded the principle to both reflection and refraction, and the principle was later stated in this form by Pierre de Fermat in 1662; the most modern form is that the path is at an extremum.
- A standalone fountain that operates under self-contained hydrostatic energy. (Heron's fountain)
- A programmable cart that was powered by a falling weight. The "program" consisted of strings wrapped around the drive axle.[12]

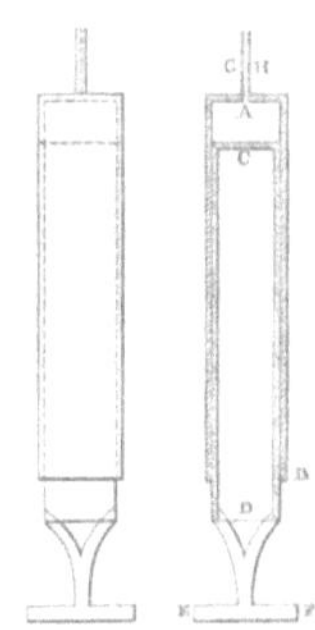
The syringe

Mathematics

Hero described a method of iteratively computing the square root.[13] Today, though, his name is most closely associated with Heron's Formula for finding the area of a triangle from its side lengths.

The imaginary number, or imaginary unit, is also noted to have been first observed by Hero while calculating the volume of a pyramidal frustum.[14]

Bibliography

The most comprehensive edition of Hero's works was published in 5 volumes in Leipzig by the publishing house Teubner in 1903.

Works known to be written by Hero:

- *Pneumatica*, a description of machines working on air, steam or water pressure, including the *hydraulis* or water organ.[15]
- *Automata*, a description of machines which enable wonders in temples by mechanical or pneumatical means (e.g. automatic opening or closing of temple doors, statues that pour wine, etc.). See Automaton.
- *Mechanica*, preserved only in Arabic, written for architects, containing means to lift heavy objects.
- *Metrica*, a description of how to calculate surfaces and volumes of diverse objects.
- *On the Dioptra*, a collection of methods to measure lengths. In this work the odometer and the dioptra, an apparatus which resembles the theodolite, are described.
- *Belopoeica*, a description of war machines.
- *Catoptrica*, about the progression of light, reflection and the use of mirrors.

Works which have sometimes been attributed to Hero, but are now thought to have most likely been written by someone else:[16]

- *Geometria*, a collection of equations based on the first chapter of *Metrica*.
- *Stereometrica*, examples of three dimensional calculations based on the second chapter of *Metrica*.
- *Mensurae*, tools which can be used to conduct measurements based on *Stereometrica* and *Metrica*.
- *Cheiroballistra*, about catapults.
- *Definitiones*, containing definitions of terms for geometry.

Works which are preserved only in fragments:

- *Geodesia*
- *Geoponica*

Media

A 2007 The History Channel television show *Ancient Discoveries* includes recreations of most of Heron's devices.

A 2008 The History Channel television show *Ancient Discoveries* - "Ancient New York" includes a short recreation of a fountain device that made water flow uphill.

A 1979 Soviet animated short film focuses on Heron's invention of the *aeolipile*, showing him as a plain craftsman who invented the turbine accidentally.[17]

See also

- Heron's formula
- Heronian triangle
- Heronian mean
- Heron's engine

References

[1] Research Machines plc. (2004). *The Hutchinson dictionary of scientific biography*. Abingdon, Oxon: Helicon Publishing. pp. 546. "**Hero of Alexandria (lived c. *AD* 60)** Greek mathematician and engineer, the greatest experimentalist of antiquity"

[2] Boyer (1968 [1991]). "Greek Trigonometry and Mensuration". *A History of Mathematics*. pp. 171–172. "At least from the days of Alexander the Great to the close of the classical world, there undoubtedly was much intercommunication between Greece and Mesopotamia, and it seems to be clear that the Babylonian arithmetic and algebraic geometry continued to exert considerable influence in the Hellenistic world. This aspect of mathematics, for example, appears so strongly in Heron of Alexandria (fl. ca. A.D. 100) that Heron once was thought to be Egyptian

or Phoenician rather than Greek. Now it is thought that Heron portrays a type of mathematics that had long been present in Greece but does not find a representative among the great figures - except perhaps as betrayed by Ptolemy in the *Tetrabiblos*."

[3] Gregory A Tokaty (1994). *A History and Philosophy of Fluid Mechanics*. Courier Dover Publications. pp. 26. ISBN 0-486-68103-3.
Grolier Incorporated (1989). *Academic American Encyclopedia*. Grolier University of Michigan. pp. 144. ISBN 0-7172-2024-9.
Encyclopædia Britannica Online - Heron of Alexandria (http://www.britannica.com/eb/article-9040189/Heron-of-Alexandria/)
Israel Shatzman, Michael Avi-Yonah (1975). *Illustrated Encyclopedia of the Classical World*. Harper and Row. pp. 234. ISBN 0-06-010178-4.
Gillian Clements (2005). *The Picture History of Great Inventors*. frances lincoln ltd. pp. 13. ISBN 0-7112-1605-3.
Enc. Britannica 2007, "Heron of Alexandria"

[4] Marie Boas, "Hero's Pneumatica: A Study of Its Transmission and Influence", Isis, Vol. 40, No. 1 (Feb., 1949), p. 38 and supra

[5] A.G. Drachmann, "Heron's Windmill", *Centaurus*, 7 (1961), pp. 145-151

[6] Dietrich Lohrmann, "Von der östlichen zur westlichen Windmühle", *Archiv für Kulturgeschichte*, Vol. 77, Issue 1 (1995), pp.1-30 (10f.)

[7] Kelly, Kevin (1994). *Out of control: the new biology of machines, social systems and the economic world*. Boston: Addison-Wesley. ISBN 0-201-48340-8.

[8] Hero of Alexandria; Bennet Woodcroft (trans.) (1851). "Temple Doors opened by Fire on an Altar" (http://www.history.rochester.edu/steam/hero/section37.html). *Pneumatics of Hero of Alexandria*. London: Taylor Walton and Maberly (online edition from University of Rochester, Rochester, NY). . Retrieved 2008-04-23.

[9] for example: Mokyr, Joel (2001). *Twenty-five centuries of technological change*. London: Routledge. p. 11. ISBN 0-415-26931-8. "Among the devices credited to Hero are the aeolipile, a working steam engine used to open temple doors" and Wood, Chris M.; McDonald, D. Gordon (1997). "History of propulsion devices and turbo machines". *Global Warming*. Cambridge, England: Cambridge University Press. p. 3. ISBN 0-521-49532-6. "Two exhaust nozzles...were used to direct the steam with high velocity and rotate the sphere...By attaching ropes to the axial shaft Heron used the developed power to perform tasks such as opening temple doors"

[10] Humphrey, John W.; John P. Oleson, Andrew N. Sherwood (1998). *Greek and Roman technology: A Sourcebook. Annotated translations of Greek and Latin texts and documents*. Routledge Sourcebooks for the Ancient World. London and New York: Routledge. ISBN 978-0-415-06137-7., pp.66–67

[11] Hero of Alexandria; Bennet Woodcroft (trans.) (1851). "No. 57. Description of a Syringe.". Pneumatics of Hero of Alexandria. London: Taylor Walton and Maberly (online edition from University of Rochester, Rochester, NY). (http://www.history.rochester.edu/steam/hero/section57.html.) Retrieved 2010 January 27

[12] Noel Sharkey (July 4, 2007), *A programmable robot from 60 AD* (http://www.newscientist.com/blog/technology/2007/07/programmable-robot-from-60ad.html), **2611**, New Scientist,

[13] Heath, Thomas (1921). *A History of Greek Mathematics, Vol. 2* (http://books.google.com/books?id=LOA5AAAAMAAJ&pg=PR323). Oxford: Clarendon Press. pp. 323–324. .

[14] Nahin, Paul."An Imaginary Tale: The story of [the square root of minus one]. Princeton University Press. 1998"

[15] Jamies W. McKinnon. "Hero of Alexandria and Hydraulis", *Grove Music Online*, ed. L. Macy (accessed January 17 2007), grovemusic.com (http://www.grovemusic.com/) (subscription access).

[16] O'Connor, J.J. and E.F. Robertson. "Heron biography" (http://www-history.mcs.st-andrews.ac.uk/history/Biographies/Heron.html). *The MacTutor History of Mathematics archive*. . Retrieved 2006-06-18.

[17] animator.ru (http://animator.ru/db/?ver=eng&p=show_film&fid=4676)

Further Reading

- Drachmann, Aage Gerhardt (1963). *The Mechanical Technology of Greek and Roman Antiquity: A Study of the Literary Sources*. Madison, WI: University of Wisconsin Press.
- Landels, J.G. (2000). *Engineering in the ancient world* (2nd ed.). Berkeley: University of California Press. ISBN 0-520-22782-4.
- Marsden, E.W. (1969). *Greek and Roman Artillery: Technical Treatises*. Oxford: Clarendon Press.
- Schellenberg, H.M.: *Anmerkungen zu Hero von Alexandria und seinem Werk über den Geschützbau*, in: Schellenberg, H.M./ Hirschmann, V.E./ Krieckhaus, A.(edd.): A Roman Miscellany. Essays in Honour of Anthony R. Birley on his Seventieth Birthday, Gdansk 2008, 92-130 (with a huge bibliography of over 300 titles and discussion of the communis opinio on Hero).

External links

- Webpage about Hero by The Technology Museum of Thessaloniki (http://www.tmth.edu.gr/en/aet/5/55.html)
- A translation of *Pneumatica* with diagrams (http://www.history.rochester.edu/steam/hero/index.html)
- Heron biography, The MacTutor History of Mathematics archive (http://www-history.mcs.st-andrews.ac.uk/Biographies/Heron.html)
- Hero of Alexandria (http://www.archive.org/stream/encyclopaediabrit13chisrich#page/378/mode/1up) in the 1911 Encyclopædia Britannica
- Online Galleries, History of Science Collections, University of Oklahoma Libraries (http://hos.ou.edu/galleries//01Ancient/HeroOfAlexandria/) High resolution images of works by Hero of Alexandria in .jpg and .tiff format.
- "Hero of Alexandria". *New International Encyclopedia.* 1905.

Nicomachus

Nicomachus (Greek: Νικόμαχος; c. 60 – c. 120 AD) was an important mathematician in the ancient world and is best known for his works *Introduction to Arithmetic* (Arithmetike eisagoge) and *Manual of Harmonics* in Greek. He was born in Gerasa, in the Roman province of Syria (now Jerash, Jordan), and was strongly influenced by Aristotle. He was a Neopythagorean, who wrote about the mystical properties of numbers.

Life

Little is known about the life of Nicomachus except that he was a Pythagorean and that he came from Gerasa. Historians consider him a Neopythagorean based on his tendency to view the numbers having mystical properties. The age in which he lived (c. 100 AD) is only known because he mentions Thrasyllus in his *Manual of Harmonics*, and because his *Introduction to Arithmetic* was apparently translated into Latin in the mid 2nd century by Apuleius.[1] His *Manual of Harmonics* was addressed to a lady of noble birth, at whose request Nicomachus wrote the book, which suggests that he was a respected scholar of some status.[1] He mentions his intent to write a more advanced work, and how the journeys he frequently undertakes leave him short of time.[1]

Works

Introduction to Arithmetic

Introduction to Arithmetic (Greek: Ἀριθμητικὴ εἰσαγωγή), the lesser work on arithmetic. As a Neo-Pythagorean, Nicomachus was often more interested in the mystical properties of numbers rather than their mathematical properties. He distinguishes between the wholly conceptual immaterial number, which he regards as the 'divine number', and the numbers which measure material things, the 'scientific' number.[1] He writes extensively on numbers, especially on the significance of prime numbers and perfect numbers and argues that arithmetic is ontologically prior to the other mathematical sciences (music, geometry, and astronomy), and is their cause. Boethius' *De institutione arithmetica* is in large part a Latin translation of this work.

Manual of Harmonics

Manual of Harmonics (Greek: Ἐγχειρίδιον ἁρμονικῆς). This is the first important music theory treatise since the time of Aristoxenus and Euclid. It provides the earliest surviving record of the story of Pythagoras's epiphany outside a smithy that pitch is determined by numeric ratios. Nicomachus also gives the first in depth account of the relationship between music and the ordering of the universe via the "music of the spheres." Nicomachus's discussion of the governance of the ear and voice in understanding music unites Aristoxenian and Pythagorean concerns, normally regarded as antitheses.[2] In the midst of theoretical discussions, Nicomachus also describes the instruments of his time, also providing a valuable resource. In addition to the *Manual*, ten extracts survive from what appear to have originally been a more substantial work on music.

Lost Works

The works which are lost are:

- *Art of Arithmetic* (Greek: Τέχνη ἀριθμητική), the larger work on arithmetic, mentioned by Photius.
- A larger work on music, promised by Nicomachus himself, and apparently referred to by Eutocius in his comment on the sphere and cylinder of Archimedes.
- An *Introduction to Geometry*, referred to by Nicomachus,[3] although whether it was his work in unclear.
- *Theology of Arithmetic* (Greek: Θεολογούμενα ἀριθμητικῆς), on the Pythagorean mystical properties of numbers in two books mentioned by Photius. There is an extant work sometimes attributed to Iamblichus under this title written two centuries later which contains a great deal of material thought to have been copied or paraphrased from Nicomachus' work.
- A *Life of Pythagoras*, one of the main sources used by Porphyry and Iamblichus, for their (extant) *Lives* of Pythagoras.
- A collection of Pythagorean *dogmata*, referred to by Iamblichus.
- *On Egyptian festivals* (Greek: Περὶ ἑορτῶν Αἰγυπτίων), is mentioned by Athenaeus, but whether by this Nicomachus is uncertain.

See also

- Monad
- Dyad
- Triad
- Pentad
- Tetrad
- Decad
- Nicomachus's theorem
- Superparticular number
- Superpartient number

Notes

[1] Henrietta Midonick, (1965), *The Treasury of Mathematics, Volume 2*, pages 15-16. Penguin Books.

[2] Levin, "Nicomachus of Gerasa," *Grove Music Online*.

[3] Nicomachus, *Arithmetica*, ii. 6. 1.

References

- Flora R. Levin, "Nicomachus of Gerasa," *Grove Music Online*, visited 3 June 2007.
- Andrew Barker, editor, *Greek Musical Writings* vol 2: *Harmonic and Acoustic Theory* (Cambridge: Cambridge University Press, 1989), pp. 245–69.

External links

- Nicomachus' "Introduction to Arithmetic" (http://www.archive.org/details/NicomachusIntroToArithmetic) translated by Martin Luther D'ooge.
- O'Connor, John J.; Robertson, Edmund F., "Nicomachus" (http://www-history.mcs.st-andrews.ac.uk/Biographies/Nicomachus.html), *MacTutor History of Mathematics archive*, University of St Andrews.
- Nicomachus' "Introduction to Arithmetic" from the Million Books Project (Greek) (http://www.archive.org/details/nicomachigerasen00nicouoft)

Menelaus_of_Alexandria

Menelaus of Alexandria (c. 70–140 CE) was a Greek[1] mathematician and astronomer, the first to recognize geodesics on a curved surface as natural analogs of straight lines.

Life and Works

Although very little is known about Menelaus's life, it is supposed that he lived in Rome, where he probably moved after having spent his youth in Alexandria. He was called *Menelaus of Alexandria* by both Pappus of Alexandria and Proclus, and a conversation of his with Lucius, held in Rome, is recorded by Plutarch.

Ptolemy (2nd century CE) also mentions, in his work *Almagest* (VII.3), two astronomical observations made by Menelaus in Rome in January of the year 98. These were occultations of the stars Spica and Beta Scorpii by the moon, a few nights apart. Ptolemy used these observations to confirm precession of the equinoxes, a phenomenon that had been discovered by Hipparchus in the 2nd century BCE.

Sphaerica is the only book that has survived, in an Arabic translation. Composed of three books, it deals with the geometry of the sphere and its application in astronomical measurements and calculations. The book introduces the concept of spherical triangle (figures formed of three great circle arcs, which he named "trilaterals") and proves Menelaus' theorem on collinearity of points on the edges of a triangle (which may have been previously known) and its analog for spherical triangles. It was later translated by the sixteenth century astronomer and mathematician Francesco Maurolico.

The lunar crater Menelaus is named after him.

Bibliography

The titles of a few books by Menelaus have been preserved:

- *On the calculation of the chords in a circle*, composed of six books
- *Elements of geometry*, composed of three books, later edited by Thabit ibn Qurra
- *On the knowledge of the weights and distributions of different bodies*
- He may also have written a star catalogue.

Notes

[1] Encyclopaedia Britannica "Greek mathematician and astronomer who first conceived and defined a spherical triangle (a triangle formed by three arcs of great circles on the surface of a sphere)."

References

- Ivor Bulmer-Thomas. "Menelaus of Alexandria." *Dictionary of Scientific Biography* 9:296-302.

External links

- O'Connor, John J.; Robertson, Edmund F., "Menelaus of Alexandria" (http://www-history.mcs.st-andrews.ac.uk/Biographies/Menelaus.html), *MacTutor History of Mathematics archive*, University of St Andrews.
- *Halley's Latin Translation from the Arabic and Hebrew Versions (Vol. III)* (http://books.google.com/books?id=Is02AAAAMAAJ) at Google Books

Pappus_of_Alexandria

Pappus of Alexandria (/ˈpæpəs/; Greek Πάππος ὁ Ἀλεξανδρεύς) (c. 290 – c. 350) was one of the last great Greek mathematicians of Antiquity, known for his *Synagoge* or *Collection* (c. 340), and for Pappus's Theorem in projective geometry. Nothing is known of his life, except (from his own writings) that he had a son named Hermodorus, and was a teacher in Alexandria.[1]

Collection, his best-known work, is a compendium of mathematics in eight volumes, the bulk of which survives. It covers a wide range of topics, including geometry, recreational mathematics, doubling the cube, polygons and polyhedra.

Context

Pappus flourished in the 4th century A.D. In a period of general stagnation in mathematical studies, he stands out as a remarkable exception.[2] How far he was above his contemporaries, how little appreciated or understood by them, is shown by the absence of references to him in other Greek writers, and by the fact that his work had no effect in arresting the decay of mathematical science. In this respect the fate of Pappus strikingly resembles that of Diophantus.[2]

Dating

In his extant writings, Pappus gives no indication of the date of the authors whose treatises he makes use of, or of the time (but see below) at which he himself wrote. If we had no other information, we should only know that he was later than Ptolemy (died c. 168 AD), whom he quotes, and earlier than Proclus (born c. 411 AD), who quotes him.

The *Suda* (a 10th century Byzantine Greek encyclopedia of known inaccuracy) states that Pappus was of the same age as Theon of Alexandria, who flourished in the reign of Emperor Theodosius I (372–395 AD).[3] A different date is given by a marginal note to a late 10th century manuscript (a copy of a chronological table by the same Theon), which states, next to an entry on Emperor Diocletian (reigned 284–305 AD), that "at that time wrote Pappus".[2]

However, a real date comes from the dating of a solar eclipse mentioned by Pappus himself, when in his commentary on the *Almagest* he calculates "the place and time of conjunction which gave rise to the eclipse in Tybi in 1068 after Nabonassar". This works out as October 18, 320 AD, and so we can finally say that Pappus flourished c. 320 AD.[1]

Works

The great work of Pappus, in eight books and titled *Synagoge* or *Collection*, we possess only in an incomplete form, the first book being lost, and the rest having suffered considerably. The *Suda* enumerates other works of Pappus: *Χωρογραφια οἰκουμενική* (*Chorography oikoumenike* or *Description of the Inhabited World*), commentary on the 4 books of Ptolemy's *Almagest*, *Ποταμοὺς τοὺς ἐν Λιβύῃ* (*The Rivers in Libya*), and *Ὀνειροκριτικά* (*The Interpretation of Dreams*).[3] Pappus himself mentions another commentary of his own on the *Ἀνάλημμα* (*Analemma*) of Diodorus of Alexandria. Pappus also wrote commentaries on Euclid's *Elements* (of which fragments are preserved in Proclus and the Scholia, while that on the tenth Book has been found in an Arabic manuscript), and on Ptolemy's *Ἁρμονικά* (*Harmonika*).[2]

Collection

The characteristics of Pappus's *Collection* are that it contains an account, systematically arranged, of the most important results obtained by his predecessors, and, secondly, notes explanatory of, or extending, previous discoveries. These discoveries form, in fact, a text upon which Pappus enlarges discursively. Very valuable are the systematic introductions to the various books which set forth clearly in outline the contents and the general scope of the subjects to be treated. From these introductions we are able to judge of the style of Pappus's writing, which is excellent and even elegant the moment he is free from the shackles of mathematical formulae and expressions. At the same time, his characteristic exactness makes his collection a most admirable substitute for the texts of the many valuable treatises of earlier mathematicians of which time has deprived us. We proceed to summarize briefly the contents of that portion of the *Collection* which has survived, mentioning separately certain propositions which seem to be among the most important.

We can only conjecture that the lost **Book I**, like Book II, was concerned with arithmetic, Book III being clearly introduced as beginning a new subject.

The whole of **Book II** (the former part of which is lost, the existing fragment beginning in the middle of the 14th proposition) discusses a method of multiplication from an unnamed book by Apollonius of Perga. The final propositions deal with multiplying together the numerical values of Greek letters in two lines of poetry, producing two huge numbers approximately equal to $2*10^{54}$ and $2*10^{38}$.[4]

Book III contains geometrical problems, plane and solid. It may be divided into five sections:

1. On the famous problem of finding two mean proportionals between two given lines, which arose from that of duplicating the cube, reduced by Hippocrates of Chios to the former. Pappus gives several solutions of this problem, including a method of making successive approximations to the solution, the significance of which he apparently failed to appreciate; he adds his own solution of the more general problem of finding geometrically the side of a cube whose content is in any given ratio to that of a given one.
2. On the arithmetic, geometric and harmonic means between two straight lines, and the problem of representing all three in one and the same geometrical figure. This serves as an introduction to a general theory of means, of which Pappus distinguishes ten kinds, and gives a table representing examples of each in whole numbers.
3. On a curious problem suggested by Euclid I.21.
4. On the inscribing of each of the five regular polyhedra in a sphere.
5. An addition by a later writer on another solution of the first problem of the book.

Of **Book IV** the title and preface have been lost, so that the program has to be gathered from the book itself. At the beginning is the well-known generalization of Euclid I.47, then follow various theorems on the circle, leading up to the problem of the construction of a circle which shall circumscribe three given circles, touching each other two and two. This and several other propositions on contact, e.g. cases of circles touching one another and inscribed in the figure made of three semicircles and known as arbelos ("shoemakers knife") form the first division of the book; Pappus turns then to a consideration of certain properties of Archimedes's spiral, the conchoid of Nicomedes (already mentioned in Book I as supplying a method of doubling the cube), and the curve discovered most probably by Hippias of Elis about 420 B.C., and known by the name, τετραγωνισμός, or quadratrix. Proposition 30 describes the construction of a curve of double curvature called by Pappus the helix on a sphere; it is described by a point moving uniformly along the arc of a great circle, which itself turns about its diameter uniformly, the point describing a quadrant and the great circle a complete revolution in the same time. The area of the surface included between this curve and its base is found – the first known instance of a quadrature of a curved surface. The rest of the book treats of the trisection of an angle, and the solution of more general problems of the same kind by means of the quadratrix and spiral. In one solution of the former problem is the first recorded use of the property of a conic (a hyperbola) with reference to the focus and directrix.

In **Book V**, after an interesting preface concerning regular polygons, and containing remarks upon the hexagonal form of the cells of honeycombs, Pappus addresses himself to the comparison of the areas of different plane figures which have all the same perimeter (following Zenodorus's treatise on this subject), and of the volumes of different solid figures which have all the same superficial area, and, lastly, a comparison of the five regular solids of Plato. Incidentally Pappus describes the thirteen other polyhedra bounded by equilateral and equiangular but not similar polygons, discovered by Archimedes, and finds, by a method recalling that of Archimedes, the surface and volume of a sphere.

According to the preface, **Book VI** is intended to resolve difficulties occurring in the so-called "lesser astronomical works" (μίκρός ἀστρονομούμενος), i.e. works other than the *Almagest*. It accordingly comments on the *Sphaerica* of Theodosius, the *Moving Sphere* of Autolycus, Theodosius's book on *Day and Night*, the treatise of Aristarchus *On the Size and Distances of the Sun and Moon*, and Euclid's *Optics and Phaenomena*.

Book VII

Since Michel Chasles cited this book of Pappus in his history of geometric methods,[5] it has become the object of considerable attention.

The preface of **Book VII** explains the terms analysis and synthesis, and the distinction between theorem and problem. Pappus then enumerates works of Euclid, Apollonius, Aristaeus and Eratosthenes, thirty-three books in all, the substance of which he intends to give, with the lemmas necessary for their elucidation. With the mention of the *Porisms* of Euclid we have an account of the relation of porism to theorem and problem. In the same preface is included (a) the famous problem known by Pappus's name, often enunciated thus: Having given a number of straight lines, to find the geometric locus of a point such that the lengths of the perpendiculars upon, or (more generally) the lines drawn from it obliquely at given inclinations to, the given lines satisfy the condition that the product of certain of them may bear a constant ratio to the product of the remaining ones; (Pappus does not express it in this form but by means of composition of ratios, saying that if the ratio is given which is compounded of the ratios of pairs one of one set and one of another of the lines so drawn, and of the ratio of the odd one, if any, to a given straight line, the point will lie on a curve given in position); (b) the theorems which were rediscovered by and named after Paul Guldin, but appear to have been discovered by Pappus himself.

Book VII contains also

1. under the head of the *De Sectione Determinata* of Apollonius, lemmas which, closely examined, are seen to be cases of the involution of six points;
2. important lemmas on the *Porisms* of Euclid;

3. a lemma upon the *Surface Loci* of Euclid which states that the locus of a point such that its distance from a given point bears a constant ratio to its distance from a given straight line is a conic, and is followed by proofs that the conic is a parabola, ellipse, or hyperbola according as the constant ratio is equal to, less than or greater than 1 (the first recorded proofs of the properties, which do not appear in Apollonius).

Chasles citation of Pappus was repeated by Wilhelm Blaschke[6] and Dirk Struik.[7] In Cambridge, England, John J. Milne[8] gave readers the benefit of his reading of Pappus. In 1985 Alexander Jones wrote his thesis at Brown University on the subject. A revised form of his translation and commentary was published by Springer-Verlag the following year. Jones succeeds in showing how Pappus manipulated the complete quadrangle, used the relation of projective harmonic conjugates, and displayed an awareness of cross-ratios of points and lines. Furthermore, the concept of pole and polar is revealed as a lemma in Book VII.

Book VIII

Lastly, **Book VIII** principally treats mechanics, the properties of the center of gravity, and some mechanical powers. Interspersed are some propositions on pure geometry. Proposition 14 shows how to draw an ellipse through five given points, and Prop. 15 gives a simple construction for the axes of an ellipse when a pair of conjugate diameters are given.

Theorems

Although **Pappus's Theorem** usually refers to Pappus's hexagon theorem, it may also refer to Pappus's centroid theorem.

He also gives his name to the Pappus chain, and to the Pappus configuration and Pappus graph arising from his hexagon theorem.

Notes

[1] Pierre Dedron, J. Itard (1959) *Mathematics And Mathematicians*, Vol. 1, p.149 (trans. Judith V. Field) (Transworld Student Library, 1974)

[2] Heath, Thomas Little (1911). "Pappus of Alexandria" (http://www.archive.org/stream/encyclopdiabri20chis#page/740/mode/2up). In Chisholm, Hugh. *Encyclopædia Britannica* (11th ed.). Cambridge University Press. pp. 740-741. .

[3] Whitehead, David, ed. "Suda On Line - Pappos" (http://www.stoa.org/sol-bin//search.pl?search_method=QUERY&login=&enlogin=&searchstr=pi,265&field=adlerhw_gr&db=REAL). Suda On Line and the Stoa Consortium. . Retrieved 2012-07-11. "Alexandrian, philosopher, born in the time of the elder emperor Theodosius, when the philosopher Theon also flourished, the one who wrote about Ptolemy's Canon. His books are Description of the Inhabited World; Commentary on the 4 Books of the Great Syntaxis of Ptolemy; The Rivers in Libya; The Interpretation of Dreams."

[4] Pappus of Alexandria, trans. into Latin by Friedrich Hultsch. *Pappi Alexandrini collectionis quae supersunt* (http://archive.org/details/pappialexandrin01hultgoog). Apud Weidmannos, 1877, pp. 19-29.

[5] Michel Chasles (1837) *Aperçu historique sur l'origine et le développement des méthodes en géométrie*, especially page 302; see also pages 12, 78, and 518.

[6] Wilhelm Blaschke (1948) *Projektiva Geometrie*, page 140

[7] Dirk Struik (1953) *Lectures in Analytic and Projective Geometry*, page 19, Addison-Wesley

[8] Milne (1911) *Elementary Treatise on Cross-Ratio Geometry*

References

- *Pappus of Alexandria* Encyclopædia Britannica.
- *The Hutchinson Dictionary of Scientific Biography* (Helicon Publishing, 2004) "Pappus of Alexandria (lived c. AD 200-350) Greek mathematician, astronomer, and geographer whose chief importance lies in his commentaries on the mathematical work of his predecessors."
- Alexander Jones (1986) *Book 7 of the Collection*, part 1: introduction, text, translation ISBN 0-387-96257-3, part 2: commentary, index, figures ISBN 3-540-96257-3, Springer-Verlag .
- John J. Milne (1911) An Elementary Treatise on Cross-Ratio Geometry with Historical Notes (http://books.google.com/books?id=BOk8AAAAIAAJ&pg=PA11&dq=Milne+Cross-ratio&hl=en&ei=_rJITaCeH4W0sAPj8fTfCg&sa=X&oi=book_result&ct=result&resnum=1&ved=0CCYQ6AEwAA#v=onepage&q&f=false), Cambridge University Press.
- Paul Ver Eecke (1933) *Pappus d'Alexandrie: La Collection Mathématique avec une Introduction et des Notes*,2 volumes, Fondation Universitaire de Belgique, Paris: Albert Blanchard.
- This article incorporates text from a publication now in the public domain: Chisholm, Hugh, ed. (1911). *Encyclopædia Britannica* (11th ed.). Cambridge University Press.

External links

- Pappos (http://www.hs-augsburg.de/~harsch/graeca/Chronologia/S_post04/Pappos/pap_intr.html) (Bibliotheca Augustana)
- O'Connor, John J.; Robertson, Edmund F., "Pappus of Alexandria" (http://www-history.mcs.st-andrews.ac.uk/Biographies/Pappus.html), *MacTutor History of Mathematics archive*, University of St Andrews.
- "Pappus" in The Columbia Electronic Encyclopedia, Sixth Edition (http://www.answers.com/topic/pappus-1) at Answer.com.
- Pappus's Theorem (http://www.mathpages.com/home/kmath542/kmath542.htm) at MathPages
- Pappus's work on the Isoperimetric Problem (http://mathdl.maa.org/convergence/1/?pa=content&sa=viewDocument&nodeId=1186&bodyId=1319) at Convergence (http://mathdl.maa.org/convergence/1/)

Pythagoras

Bust of Pythagoras of Samos in the Capitoline Museums, Rome	
Born	c. 570 BCSamos
Died	c. 495 BC (aged around 75)Metapontum
Era	Ancient philosophy
Region	Western philosophy
School	Pythagoreanism
Main interests	Metaphysics, Music, Mathematics, Ethics, Politics
Notable ideas	Musica universalis, Golden ratio, Pythagorean tuning, Pythagorean theorem

Pythagoras of Samos (Ancient Greek: Πυθαγόρας ὁ Σάμιος [Πυθαγόρης in Ionian Greek] *Pythagóras ho Sámios* "Pythagoras the Samian", or simply Πυθαγόρας; b. about 570 – d. about 495 BC)[1] [2] was an Ionian Greek philosopher, mathematician, and founder of the religious movement called Pythagoreanism. Most of the information about Pythagoras was written down centuries after he lived, so very little reliable information is known about him. He was born on the island of Samos, and might have travelled widely in his youth, visiting Egypt and other places seeking knowledge. Around 530 BC, he moved to Croton, a Greek colony in southern Italy, and there set up a religious sect. His followers pursued the religious rites and practices developed by Pythagoras, and studied his philosophical theories. The society took an active role in the politics of Croton, but this eventually led to their downfall. The Pythagorean meeting-places were burned, and Pythagoras was forced to flee the city. He is said to have ended his days in Metapontum.

Pythagoras made influential contributions to philosophy and religious teaching in the late 6th century BC. He is often revered as a great mathematician, mystic and scientist, but he is best known for the Pythagorean theorem which bears his name. However, because legend and obfuscation cloud his work even more than that of the other pre-Socratic philosophers, one can give only a tentative account of his teachings, and some have questioned whether he contributed much to mathematics and natural philosophy. Many of the accomplishments credited to Pythagoras may actually have been accomplishments of his colleagues and successors. Whether or not his disciples believed that everything was related to mathematics and that numbers were the ultimate reality is unknown. It was said that he was the first man to call himself a philosopher, or lover of wisdom,[3] and Pythagorean ideas exercised a marked influence on Plato, and through him, all of Western philosophy.

Biographical sources

Accurate facts about the life of Pythagoras are so few, and most information concerning him is of so late a date, and so untrustworthy, that it is impossible to provide more than a vague outline of his life. The lack of information by contemporary writers, together with the secrecy which surrounded the Pythagorean brotherhood, meant that invention took the place of facts. The stories which were created were eagerly sought by the Neoplatonist writers who provide most of the details about Pythagoras, but who were uncritical concerning anything which related to the gods or which was considered divine.[4] Thus many myths were created – such as that Apollo was his father; that Pythagoras gleamed with a supernatural brightness; that he had a golden thigh; that Abaris came flying to him on a golden arrow; that he was seen in different places at one and the same time.[5] With the exception of a few remarks by Xenophanes, Heraclitus, Herodotus, Plato, Aristotle, and Isocrates, we are mainly dependent on Diogenes Laërtius, Porphyry, and Iamblichus for the biographical details. Aristotle had written a separate work on the Pythagoreans, which unfortunately has not survived.[6] His disciples Dicaearchus, Aristoxenus, and Heraclides

Ponticus had written on the same subject. These writers, late as they are, were among the best sources from whom Porphyry and Iamblichus drew, besides the legendary accounts and their own inventions. Hence historians are often reduced to considering the statements based on their inherent probability, but even then, if all the credible stories concerning Pythagoras were supposed true, his range of activity would be impossibly vast.[7]

Life

Bust of Pythagoras, Vatican

Herodotus, Isocrates, and other early writers all agree that Pythagoras was born on Samos, the Greek island in the eastern Aegean, and we also learn that Pythagoras was the son of Mnesarchus.[8] His father was a gem-engraver or a merchant. His name led him to be associated with Pythian Apollo; Aristippus explained his name by saying, "He spoke (*agor-*) the truth no less than did the Pythian (*Pyth-*)," and Iamblichus tells the story that the Pythia prophesied that his pregnant mother would give birth to a man supremely beautiful, wise, and beneficial to humankind.[9] A late source gives his mother's name as Pythais.[10] As to the date of his birth, Aristoxenus stated that Pythagoras left Samos in the reign of Polycrates, at the age of 40, which would give a date of birth around 570 BC.[11]

It was natural for the ancient biographers to inquire as to the origins of Pythagoras' remarkable system. In the absence of reliable information, however, a huge range of teachers were assigned to Pythagoras. Some made his training almost entirely Greek, others exclusively Egyptian and Oriental. We find mentioned as his instructors Creophylus,[12] Hermodamas of Samos,[13] Bias,[12] Thales,[12] Anaximander,[14] and Pherecydes of Syros.[15] He is said too, to have been taught by a Delphic priestess named Themistoclea, who introduced him to the principles of ethics.[16] [17] The Egyptians are said to have taught him geometry, the Phoenicians arithmetic, the Chaldeans astronomy, the Magians the principles of religion and practical maxims for the conduct of life.[18] Of the various claims regarding his Greek teachers, Pherecydes is mentioned most often.

Diogenes Laertius reported that Pythagoras had undertaken extensive travels, and had visited not only Egypt, but Arabia, Phoenicia, Judaea, Babylon, and even India, for the purpose of collecting all available knowledge, and especially to learn information concerning the secret or mystic cults of the gods.[19] Plutarch asserted in his book *On Isis and Osiris* that during his visit to Egypt, Pythagoras received instruction from the Egyptian priest Oenuphis of Heliopolis.[20] Other ancient writers asserted his visit to Egypt.[21] Enough of Egypt was known to attract the curiosity of an inquiring Greek, and contact between Samos and other parts of Greece with Egypt is mentioned.[22]

It is not easy to say how much Pythagoras learned from the Egyptian priests, or indeed, whether he learned anything at all from them. There was nothing in the symbolism which the Pythagoreans adopted which showed the distinct traces of Egypt. The secret religious rites of the Pythagoreans exhibited nothing but what might have been adopted in the spirit of Greek religion, by those who knew nothing of Egyptian mysteries. The philosophy and the institutions of Pythagoras might easily have been developed by a Greek mind exposed to the ordinary influences of the age. Even the ancient authorities note the similarities between the religious and ascetic peculiarities of Pythagoras with the Orphic or Cretan mysteries,[23] or the Delphic oracle.[24]

There is little direct evidence as to the kind and amount of knowledge which Pythagoras acquired, or as to his definite philosophical views. Everything of the kind mentioned by Plato and Aristotle is attributed not to Pythagoras, but to the Pythagoreans. Heraclitus stated that he was a man of extensive learning;[25] and Xenophanes claimed that he believed in the transmigration of souls.[26] Xenophanes mentions the story of his interceding on behalf of a dog that was being beaten, professing to recognise in its cries the voice of a departed friend. Pythagoras is supposed to have claimed that he had been Euphorbus, the son of Panthus, in the Trojan war, as well as various other characters, a tradesman, a courtesan, etc.[27] In his book *The Life of Apollonius of Tyana*, Philostratus wrote that Pythagoras knew not only who he was himself, but also who he had been.[28]

Many mathematical and scientific discoveries were attributed to Pythagoras, including his famous theorem,[29] as well as discoveries in the field of music,[30] astronomy,[31] and medicine.[32] But it was the religious element which made the profoundest impression upon his contemporaries. Thus the people of Croton were supposed to have identified him with the Hyperborean Apollo,[33] and he was said to have practised divination and prophecy.[34] In the visits to various places in Greece – Delos, Sparta, Phlius, Crete, etc. which are ascribed to him, he usually appears either in his religious or priestly guise, or else as a lawgiver.[35]

After his travels, Pythagoras moved (around 530 BC) to Croton, in Italy (Magna Graecia). Possibly the tyranny of Polycrates in Samos made it difficult for him to achieve his schemes there. His later admirers claimed that Pythagoras was so overburdened with public duties in Samos, because of the high estimation in which he was held by his fellow-citizens, that he moved to Croton.[36] On his arrival in Croton, he quickly attained extensive influence, and many people began to follow him. Later biographers tell fantastical stories of the effects of his eloquent speech in leading the people of Croton to abandon their luxurious and corrupt way of life and devote themselves to the purer system which he came to introduce.[37]

His followers established a select brotherhood or club for the purpose of pursuing the religious and ascetic practices developed by their master. The accounts agree that what was done and taught among the members was kept a profound secret. The esoteric teachings may have concerned the secret religious doctrines and usages, which were undoubtedly prominent in the Pythagorean system, and may have been connected with the worship of Apollo.[38] Temperance of all kinds seems to have been strictly urged. There is disagreement among the biographers as to whether Pythagoras forbade all animal food,[39] or only certain types.[40] The club was in practice at once "a philosophical school, a religious brotherhood, and a political association."[41]

Pythagoras, depicted on a 3rd-century coin

Such an aristocratic and exclusive club could easily have made many people in Croton jealous and hostile, and this seems to have led to its destruction. The circumstances, however, are uncertain. Conflict seems to have broken out between the towns of Sybaris and Croton. The forces of Croton were headed by the Pythagorean Milo, and it is likely that the members of the brotherhood took a prominent part. After the decisive victory by Croton, a proposal for establishing a more democratic constitution, was unsuccessfully resisted by the Pythagoreans. Their enemies, headed by Cylon and Ninon, the former of whom is said to have been irritated by his exclusion from the brotherhood, roused the populace against them. An attack was made upon them while assembled either in the house of Milo, or in some other meeting-place. The building was set on fire, and many of the assembled members perished; only the younger and more active escaping.[42] Similar commotions ensued in the other cities of Magna Graecia in which Pythagorean clubs had been formed.

As an active and organised brotherhood the Pythagorean order was everywhere suppressed, and did not again revive. Still the Pythagoreans continued to exist as a sect, the members of which kept up among themselves their religious observances and scientific pursuits, while individuals, as in the case of Archytas, acquired now and then great political influence. Concerning the fate of Pythagoras himself, the accounts varied. Some say that he perished in the temple with his disciples,[43] others that he fled first to Tarentum, and that, being driven from there, he escaped to Metapontum, and there starved himself to death.[44] His tomb was shown at Metapontum in the time of Cicero.[45]

According to some accounts Pythagoras married Theano, a lady of Croton. Their children are variously stated to have included a son, Telauges, and three daughters, Damo, Arignote, and Myia.

Writings

No texts by Pythagoras are known to have survived, although forgeries under his name — a few of which remain extant — did circulate in antiquity. Critical ancient sources like Aristotle and Aristoxenus cast doubt on these writings. Ancient Pythagoreans usually quoted their master's doctrines with the phrase *autos ephe* ("he himself said") — emphasizing the essentially oral nature of his teaching.

Mathematics

> The so-called Pythagoreans, who were the first to take up mathematics, not only advanced this subject, but saturated with it, they fancied that the principles of mathematics were the principles of all things.
>
> —Aristotle, *Metaphysics 1–5* , cc. 350 BC

Pythagorean theorem

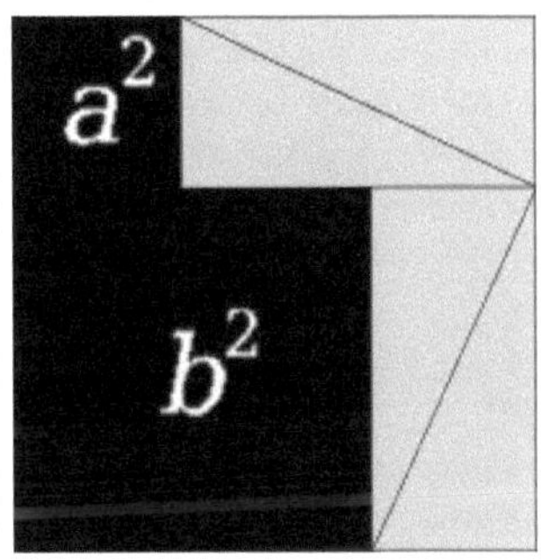

A visual proof of the Pythagorean theorem

Since the fourth century AD, Pythagoras has commonly been given credit for discovering the Pythagorean theorem, a theorem in geometry that states that in a right-angled triangle the area of the square on the hypotenuse (the side opposite the right angle) is equal to the sum of the areas of the squares of the other two sides—that is, .

While the theorem that now bears his name was known and previously utilized by the Babylonians and Indians, he, or his students, are often said to have constructed the first proof. It must, however, be stressed that the way in which the Babylonians handled Pythagorean numbers implies that they knew that the principle was generally applicable, and knew some kind of proof, which has not yet been found in the (still largely unpublished) cuneiform sources.[46] Because of the secretive nature of his school and the custom of its students to attribute everything to their teacher, there is no evidence that Pythagoras himself worked on or proved this theorem. For that matter, there is no evidence that he worked on any mathematical or meta-mathematical problems. Some attribute it as a carefully constructed myth by followers of Plato over two centuries after the death of Pythagoras, mainly to bolster the case for Platonic meta-physics, which resonate well with the ideas they attributed to Pythagoras. This attribution has stuck down the centuries up to modern times.[47] The earliest known mention of Pythagoras's name in connection with the theorem occurred five centuries after his death, in the writings of Cicero and Plutarch.

Musical theories and investigations

According to legend, the way Pythagoras discovered that musical notes could be translated into mathematical equations was when one day he passed blacksmiths at work, and thought that the sounds emanating from their anvils being hit were beautiful and harmonious and decided that whatever scientific law caused this to happen must be mathematical and could be applied to music. He went to the blacksmiths to learn how this had happened by looking at their tools, he discovered that it was because the hammers were "simple ratios of each other, one was half the size of the first, another was 2/3 the size, and so on."

This legend has since proven to be false by virtue of the fact that these ratios are only relevant to string length (such as the string of a monochord), and not to hammer weight.[48] [49] However, it may be that Pythagoras was indeed responsible for discovering these properties of string length.

Pythagoreans elaborated on a theory of numbers, the exact meaning of which is still debated among scholars. Another belief attributed to Pythagoras was that of the "harmony of the spheres". Thus the planets and stars moved according to mathematical equations, which corresponded to musical notes and thus produced a symphony.[50]

Tetractys

Pythagoras was also credited with devising the tetractys, the triangular figure of four rows, which add up to the perfect number, ten. As a mystical symbol, it was very important to the worship of the Pythagoreans, who would swear oaths by it:

> And the inventions were so admirable, and so divinised by those who understood them, that the members used them as forms of oath: "By him who handed to our generation the *tetractys*, source of the roots of ever-flowing nature."
>
> —Iamblichus, *Vit. Pyth.*, 29

Religion and science

Pythagoras' religious and scientific views were, in his opinion, inseparably interconnected. Religiously, Pythagoras was a believer of metempsychosis. He believed in transmigration, or the reincarnation of the soul again and again into the bodies of humans, animals, or vegetables until it became immortal. His ideas of reincarnation were influenced by ancient Greek religion. Heraclides Ponticus reports the story that Pythagoras claimed that he had lived four previous lives that he could remember in detail.[51] One of his past lives, as reported by Aulus Gellius, was as a beautiful courtesan.[52] According to Xenophanes, Pythagoras heard the cry of his dead friend in the bark of a dog.[53]

Lore

Pythagoras became the subject of elaborate legends surrounding his historic persona. Aristotle described Pythagoras as a wonder-worker and somewhat of a supernatural figure, attributing to him such aspects as a golden thigh, which was a sign of divinity. According to Muslim tradition, Pythagoras was said to have been initiated by Hermes (Egyptian Thoth).[54] According to Aristotle and others' accounts, some ancients believed that he had the ability to travel through space and time, and to communicate with animals and plants.[55] An extract from Brewer's Dictionary of Phrase and Fable's entry entitled "Golden Thigh":

> *Pythagoras is said to have had a golden thigh, which he showed to Abaris, the Hyperborean priest, and exhibited in the Olympic games.*[56]

Another legend describes his writing on the moon:

> *Pythagoras asserted he could write on the moon. His plan of operation was to write on a looking-glass in blood, and place it opposite the moon, when the inscription would appear photographed or reflected on the moon's disc.*[57]

Pythagoreans

Pythagoras, the man in the center with the book, teaching music, in The School of Athens by Raphael

Both Plato and Isocrates affirm that, above all else, Pythagoras was famous for leaving behind him a way of life.[58] Both Iamblichus and Porphyry give detailed accounts of the organisation of the school, although the primary interest of both writers is not historical accuracy, but rather to present Pythagoras as a divine figure, sent by the gods to benefit humankind.[59]

Pythagoras set up an organization which was in some ways a school, in some ways a brotherhood (and here it should be noted that sources indicate that as well as men there were many women among the adherents of Pythagoras),[60] and in some ways a monastery. It was based upon the religious teachings of Pythagoras and was very secretive. The adherents were bound by a vow to Pythagoras and each other, for the purpose of pursuing the religious and ascetic observances, and of studying his religious and philosophical theories. The claim that they put all their property into a common stock is perhaps only a later inference from certain Pythagorean maxims and practices.[61]

As to the internal arrangements of the sect, we are informed that what was done and taught among the members was kept a profound secret towards all. Porphyry stated that this silence was "of no ordinary kind." Candidates had to pass through a period of probation, in which their powers of maintaining silence (*echemythia*) were especially tested, as well as their general temper, disposition, and mental capacity.[62] There were also gradations among the members themselves. It was an old Pythagorean maxim, that every thing was not to be told to every body.[63] Thus the Pythagoreans were divided into an inner circle called the *mathematikoi* ("learners") and an outer circle called the *akousmatikoi* ("listeners").[64] Iamblichus describes them in terms of *esoterikoi* and *exoterikoi* (or alternatively *Pythagoreioi* and *Pythagoristai*),[65] according to the degree of intimacy which they enjoyed with Pythagoras. Porphyry wrote "the *mathematikoi* learned the more detailed and exactly elaborated version of this knowledge, the *akousmatikoi* (were) those who had heard only the summary headings of his (Pythagoras's) writings, without the more exact exposition."

There were ascetic practices (many of which had, perhaps, a symbolic meaning) in the way of life of the sect.[66] Some represent Pythagoras as forbidding all animal food, advocating a plant-based diet, and prohibiting consumption of beans. This may have been due to the doctrine of metempsychosis.[67] Other authorities contradict the statement. According to Aristoxenus,[68] he allowed the use of all kinds of animal food except the flesh of oxen used for ploughing, and rams.[69] There is a similar discrepancy as to the prohibition of fish and beans.[70] But temperance of all kinds seems to have been urged. It is also stated that they had common meals, resembling the Spartan system, at which they met in companies of ten.[71]

Considerable importance seems to have been attached to music and gymnastics in the daily exercises of the disciples. Their whole discipline is represented as encouraging a lofty serenity and self-possession, of which, there were various anecdotes in antiquity.[72] Iamblichus (apparently on the authority of Aristoxenus)[73] gives a long description of the daily routine of the members, which suggests many similarities with Sparta. The members of the sect showed a devoted attachment to each other, to the exclusion of those who did not belong to their ranks.[74] There were even stories of secret symbols, by which members of the sect could recognise each other, even if they had never met before.[75]

Influence

Influence on Plato

Pythagoras, or in a broader sense, the Pythagoreans, allegedly exercised an important influence on the work of Plato. According to R. M. Hare, this influence consists of three points: (1) The platonic Republic might be related to the idea of "a tightly organized community of like-minded thinkers", like the one established by Pythagoras in Croton. (2) There is evidence that Plato possibly took from Pythagoras the idea that mathematics and, generally speaking, abstract thinking is a secure basis for philosophical thinking as well as "for substantial theses in science and morals". (3) Plato and Pythagoras shared a "mystical approach to the soul and its place in the material world". It is probable that both were influenced by Orphism.[76]

Pythagoras, depicted as a medieval scholar in the *Nuremberg Chronicle*

Aristotle claimed that the philosophy of Plato closely followed the teachings of the Pythagoreans,[77] and Cicero repeats this claim: *Platonem ferunt didicisse Pythagorea omnia* ("They say Plato learned all things Pythagorean").[78] Bertrand Russell, in his *A History of Western Philosophy*, contended that the influence of Pythagoras on Plato and others was so great that he should be considered the most influential of all Western philosophers.

Influence on esoteric groups

Pythagoras started a secret society called the *Pythagorean Brotherhood* devoted to the study of mathematics. This had a great effect on future esoteric traditions, such as Freemasonry and Rosicrucianism, both of which were scientific/mystical groups dedicated to the study of mathematics/geometry and logical reasoning as opposed to religious dogma. Both Freemasonry and Rosicrucianism have claimed to have evolved out of the Pythagorean Brotherhood. The mystical and occult qualities of Pythagorean mathematics are discussed in a chapter of Manly P. Hall's *The Secret Teachings of All Ages* entitled "Pythagorean Mathematics".[79]

See also

• Apollonius of Tyana	• Pythagorean cup
• Lute of Pythagoras	• Pythagoras tree (fractal)
• Neopythagoreanism	• Sacred geometry
• Pythagorean comma	• The golden verses of Pythagoras
• Pythagorean triple	• List of things named after Pythagoras
	• Isopsephy (Gematria)

References

[1] "The dates of his life cannot be fixed exactly, but assuming the approximate correctness of the statement of Aristoxenus (ap. Porph. *V.P.* 9) that he left Samos to escape the tyranny of Polycrates at the age of forty, we may put his birth round about 570 BC, or a few years earlier. The length of his life was variously estimated in antiquity, but it is agreed that he lived to a fairly ripe old age, and most probably he died at about seventy-five or eighty." William Keith Chambers Guthrie, (1978), *A history of Greek philosophy, Volume 1: The earlier Presocratics and the Pythagoreans*, page 173. Cambridge University Press

[2] Biographies (http://www-groups.dcs.st-and.ac.uk/history/Biographies/Pythagoras.html)

[3] Cicero, *Tusculan Disputations*, 5.3.8–9 = Heraclides Ponticus fr. 88 Wehrli, Diogenes Laërtius 1.12, 8.8, Iamblichus *VP* 58. Burkert attempted to discredit this ancient tradition, but it has been defended by C.J. De Vogel, *Pythagoras and Early Pythagoreanism* (1966), pp. 97–102, and C. Riedweg, *Pythagoras: His Life, Teaching, And Influence* (2005), p. 92.

[4] Iamblichus, *Adhort. ad Philos.* p. 324, ed. Kiessling.

[5] Comp. Herodian, iv. 94, etc.

[6] He alludes to it himself, *Met.* i. 5. p. 986. 12, ed. Bekker.

[7] This article incorporates text from a publication now in the public domain: Smith, William, ed. (1867). "article name needed". *Dictionary of Greek and Roman Biography and Mythology*.

[8] Herodotus, iv. 95, Isocrates, *Busiris*, 28–9; Later writers called him a Tyrrhenian or Phliasian, and gave Marmacus, or Demaratus, as the name of his father, Diogenes Laërtius, viii. 1; Porphyry, *Vit. Pyth.* 1, 2; Justin, xx. 4; Pausanias, ii. 13.

[9] Riedweg, Christoph (2005). *Pythagoras: His Life, Teaching and Influence*. Cornell University. pp. 5–6, 59, 73.

[10] Apollonius of Tyana ap. Porphyry, *Vit. Pyth.* 2

[11] Porphyry, *Vit. Pyth.* 9

[12] Iamblichus, *Vit. Pyth.* 9

[13] Porphyry, *Vit. Pyth.* 2, Diogenes Laërtius, viii. 2 C. Riedweg, S. Rendall (http://books.google.co.uk/books?id=A8ixyQJA7_MC&pg=PA10&lpg=PA10&dq=teachers+of+Pythagoras+instructors&source=bl&ots=R1Nf9AFGWB&sig=JKF2yU5Gx0pW8ndITfrGlhHpTLU&hl=en&sa=X&ei=jmUyT5ryK4-dOpKE9d8G&sqi=2&ved=0CGUQ6AEwCA#v=onepage&q=teachers of Pythagoras instructors&f=false) ISBN 0-8014-7452-3 Retrieved 2012-02-08

[14] Iamblichus, *Vit. Pyth.* 9; Porphyry, *Vit. Pyth.* 2

[15] Aristoxenus and others in Diogenes Laërtius, i. 118, 119; Cicero, *de Div.* i. 49

[16] Mary Ellen Waithe, *Ancient women philosophers, 600 B.C.–500 A.D.*, p. 11 (http://books.google.com.br/books?id=x7ngECDpxmMC&printsec=frontcover)

[17] Malone, John C. (30 June 2009). *Psychology: Pythagoras to present* (http://books.google.com/books?id=e6Qa6cMQj8AC&pg=PA22). MIT Press. p. 22. ISBN 978-0-262-01296-6. . Retrieved 25 October 2010.

[18] Porphyry, *Vit. Pyth.* 6

[19] Diogenes Laërtius, viii. 2; Porphyry, *Vit. Pyth.* 11, 12; Iamblichus, *Vit. Pyth.* 14, etc.

[20] Plutarch, *On Isis And Osiris* (http://penelope.uchicago.edu/Thayer/E/Roman/Texts/Plutarch/Moralia/Isis_and_Osiris*/A.html), ch. 10.

[21] Antiphon. ap. Porphyry, *Vit. Pyth.* 7; Isocrates, *Busiris*, 28–9; Cicero, *de Finibus*, v. 27; Strabo, xiv.

[22] Herodotus, ii. 134, 135, iii. 39.

[23] Iamblichus, *Vit. Pyth.* 25; Porphyry, *Vit. Pyth.* 17; Diogenes Laërtius, viii. 3

[24] Ariston. ap. Diogenes Laërtius, viii. 8, 21; Porphyry, *Vit. Pyth.* 41

[25] Diogenes Laërtius, viii. 6, ix. 1, comp. Herodotus, i. 29, ii. 49, iv. 95

[26] Diogenes Laërtius, viii. 36, comp. Aristotle, *de Anima*, i. 3; Herodotus, ii. 123.

[27] Porphyry, *Vit. Pyth.* 26; Pausanias, ii. 17; Diogenes Laërtius, viii. 5; Horace, *Od.* i. 28,1. 10

[28] Flavius Philostratus, *The Life of Apollonius of Tyana* , , trad. F. C. Conybeare, Vol. 2, London, 1912, Book VI, p. 39.

[29] Diogenes Laërtius, viii. 12 ; Plutarch, *Non posse suav. vivi sec. Ep.* p. 1094

[30] Porphyry, in *Ptol. Harm.* p. 213; Diogenes Laërtius, viii. 12

[31] Diogenes Laërtius, viii. 14 ; Pliny, *Hist. Nat.* ii. 8

[32] Diogenes Laërtius, viii. 12, 14, 32

[33] Porphyry, *Vit. Pyth.* 20; Iamblichus, *Vit. Pyth.* 31, 140; Aelian, *Varia Historia*, ii. 26; Diogenes Laërtius, viii. 36.

[34] Cicero, *de Divin.* i. 3, 46; Porphyry, *Vit. Pyth.* 29.

[35] Iamblichus, *Vit. Pyth.* 25; Porphyry, *Vit. Pyth.* 17; Diogenes Laërtius, viii. 3, 13; Cicero, *Tusc. Qu.* v. 3

[36] Iamblichus, *Vit. Pyth.* 28; Porphyry, *Vit. Pyth.* 9

[37] Porphyry, *Vit. Pyth.* 18; Iamblichus, *Vit. Pyth.* 37, etc.

[38] Aelian, *Varia Historia*, ii. 26; Diogenes Laërtius, viii. 13; Iamblichus, *Vit. Pyth.* 8, 91, 141

[39] as Empedocles did afterwards, Aristotle, *Rhet.* i. 14. § 2; Sextus Empiricus, ix. 127. This was also one of the Orphic precepts, Aristoph. *Ran.* 1032

[40] Aristo ap. Diogenes Laërtius, viii. 20; comp. Porphyry, *Vit. Pyth.* 7; Iamblichus, *Vit. Pyth.* 85, 108

[41] Thirlwall, *Hist. of Greece*, vol. ii. p. 148

[42] Iamblichus, *Vit. Pyth.* 255–259; Porphyry, *Vit. Pyth.* 54–57; Diogenes Laërtius, viii. 39; comp. Plutarch, *de Gen. Socr.* p. 583

[43] Arnob. *adv. Gentes*, i. p. 23

[44] Diogenes Laërtius, viii. 39, 40; Porphyry, *Vit. Pyth.* 56; Iamblichus, *Vit. Pyth.* 249; Plutarch, *de Stoic. Rep.* 37

[45] Cicero, *de Fin.* v. 2

[46] There are about 100,000 unpublished cuneiform sources in the British Museum alone. Babylonian knowledge of proof of the Pythagorean Theorem is discussed by J. Høyrup, 'The Pythagorean "Rule" and "Theorem" – Mirror of the Relation between Babylonian and Greek Mathematics,' in: J. Renger (red.): *Babylon. Focus mesopotamischer Geschichte, Wiege früher Gelehrsamkeit, Mythos in der Moderne* (1999).

[47] From Christoph Riedweg , *Pythagoras, His Life, Teaching and Influence*, Cornell: Cornell University Press, 2005: "Had Pythagoras and his teachings not been since the early Academy overwritten with Plato's philosophy, and had this 'palimpsest' not in the course of the Roman Empire achieved unchallenged authority among Platonists, it would be scarcely conceivable that scholars from the Middle Ages and modernity down to the present would have found the Presocratic charismatic from Samos so fascinating. In fact, as a rule it was the image of Pythagoras elaborated by Neopythagoreans and Neoplatonists that determined the idea of what was Pythagorean over the centuries."

[48] Weiss, Piero, and Richard Taruskin, eds. Music in the Western World: A History in Documents. 2nd ed. N.p.: Thomson Schirmer, 1984. 3. Print.

[49] Christensen, Thomas, ed. The Cambridge history of Western music theory. Cambridge: Cambridge University Press, 2002. 143. Print. (http://books.google.com/books?id=ioa9uW2t7AQC&pg=PA143&lpg=PA143&dq=pythagoras+hammers+myth&source=bl&ots=wtLXShWfVQ&sig=LlyPLjIoE0O25ieAj9-2h7J1Qq8&hl=en&ei=itsgTduSB8L7lwf6os3UCw&sa=X&oi=book_result&ct=result&resnum=4&ved=0CCoQ6AEwAw#v=onepage&q=pythagoras hammers myth&f=false)

[50] Christoph Riedweg, Pythagoras: His Life, Teaching and Influence, Cornell: Cornell University Press, 2005 .

[51] Diogenes Laërtius, viii. 3–4

[52] Aulus Gellius, iv. 11

[53] Diogenes Laërtius, viii. 36

[54] See Antoine Faivre, in *The Eternal Hermes* (1995)

[55] Huffman, Carl. Pythagoras (Stanford Encyclopedia of Philosophy) (http://plato.stanford.edu/entries/pythagoras/#PytWon)

[56] Brewer, E. Cobham, *Brewer's Dictionary of Phrase and Fable* (http://www.infoplease.com/dictionary/brewers/golden-thigh.html)

[57] Brewer, E. Cobham, *Brewer's Dictionary of Phrase and Fable* (http://www.infoplease.com/dictionary/brewers/pythagoras.html)

[58] Plato, *Republic*, 600a, Isocrates, *Busiris*, 28

[59] John Dillon and Jackson Hershbell, (1991), *Iamblichus, On the Pythagorean Way of Life*, page 14. Scholars Press.; D. J. O'Meara, (1989), *Pythagoras Revived. Mathematics and Philosophy in Late Antiquity*, pages 35–40. Clarendon Press.

[60] Porphyry, *Vit. Pyth.* 19

[61] comp. Cicero, *de Leg.* i. 12, *de Off.* i. 7; Diogenes Laërtius, viii. 10

[62] Aristonexus ap. Iamblichus, *Vit. Pyth.* 94

[63] Diogenes Laërtius, viii. 15; Aristonexus ap. Iamblichus, *Vit. Pyth.* 31

[64] Iamblichus, *Vit. Pyth.* 80, cf. Aulus Gellius, i. 9

[65] Iamblichus, *Vit. Pyth.* 80

[66] comp. Porphyry, *Vit. Pyth.* 32; Iamblichus, *Vit. Pyth.* 96, etc.

[67] Plutarch, *de Esu Carn.* pp. 993, 996, 997

[68] Aristoxenus ap. Diogenes Laërtius, viii. 20

[69] comp. Porphyry, *Vit. Pyth.* 7; Iamblichus, *Vit. Pyth.* 85, 108

[70] Diogenes Laërtius, viii. 19, 34; Aulus Gellius, iv. 11; Porphyry, *Vit. Pyth.* 34, *de Abst.* i. 26; Iamblichus, *Vit. Pyth.* 98

[71] Iamblichus, *Vit. Pyth.* 98; Strabo, vi.

[72] Athenaeus, xiv. 623; Aelian, *Varia Historia*, xiv. 18; Iamblichus, *Vit. Pyth.* 197

[73] Iamblichus, *Vit. Pyth.* 96–101

[74] Aristonexus ap. Iamblichus, *Vit. Pyth.* 94, 101, etc., 229, etc.; comp. the story of Damon and Phintias; Porphyry, *Vit. Pyth.* 60; Iamblichus, *Vit. Pyth.* 233, etc.

[75] Scholion ad Aristophanes, *Nub.* 611; Iamblichus, *Vit. Pyth.* 237, 238

[76] R.M. Hare, Plato in C.C.W. Taylor, R.M. Hare and Jonathan Barnes, Greek Philosophers, Socrates, Plato, and Aristotle, Oxford: Oxford University Press, 1999 (1982), 103–189, here 117–9.

[77] Metaphysics, 1.6.1 (987a)

[78] Tusc. Disput. 1.17.39.

[79] Hall, Manly *The Secrets Teaching of All Ages* Tarcher Penguin 2003 pages 191–221.

Sources

Classical secondary sources

Only a few relevant source texts deal with Pythagoras and the Pythagoreans, most are available in different translations. Other texts usually build solely on information in these works.

- Diogenes Laërtius, *Vitae philosophorum VIII* (*Lives of Eminent Philosophers*), c. 200 AD, which in turn references the lost work *Successions of Philosophers* by Alexander Polyhistor — *Life of Pythagoras*, translated by Robert Drew Hicks (1925).
- Porphyry, *Vita Pythagorae* (*Life of Pythagoras*), c. 270 AD — *Porphyry, Life of Pythagoras* (http://www.ccel.org/ccel/pearse/morefathers/files/porphyry_life_of_pythagoras_02_text.htm), translated by Kenneth Sylvan Guthrie (1920)
- Iamblichus, *De Vita Pythagorica* (*On the Pythagorean Life*), c. 300 AD — *Iamblichus, Life of Pythagoras* (http://www.completepythagoras.net/mainframeset.html), translated by Kenneth Sylvan Guthrie (1920)
- Apuleius also writes about Pythagoras in *Apologia*, c. 150 AD, including a story of him being taught by Babylonian disciples of Zoroaster
- Hierocles of Alexandria, *Golden Verses of Pythagoras*, c. 430 AD

Modern secondary sources

- Burkert, Walter. *Lore and Science in Ancient Pythagoreanism*. Harvard University Press, June 1, 1972. ISBN 0-674-53918-4
- Burnyeat, M. F. "The Truth about Pythagoras" (http://www.lrb.co.uk/v29/n04/burn02_.html). *London Review of Books*, 22 February 2007.
- Guthrie, W. K. *A History of Greek Philosophy: Earlier Presocratics and the Pythagoreans*, Cambridge University Press, 1979. ISBN 0-521-29420-7
- Kingsley, Peter. *Ancient Philosophy, Mystery, and Magic: Empedocles and the Pythagorean Tradition*. Oxford University Press, 1995.
- Hermann, Arnold. *To Think Like God: Pythagoras and Parmenides—the Origins of Philosophy*. Parmenides Publishing, 2005. ISBN 978-1-930972-00-1
- O'Meara, Dominic J. *Pythagoras Revived*. Oxford University Press, 1989. ISBN 0-19-823913-0 (paperback), ISBN 0-19-824485-1 (hardcover)

External links

- Pythagoras (http://www.bbc.co.uk/programmes/b00p693b) on *In Our Time* at the BBC. (listen now (http://www.bbc.co.uk/iplayer/console/b00p693b/In_Our_Time_Pythagoras))
- Pythagoras (http://plato.stanford.edu/entries/pythagoras) entry by Carl Huffman in the *Stanford Encyclopedia of Philosophy*
- *Pythagoras of Samos* (http://www-history.mcs.st-andrews.ac.uk/Mathematicians/Pythagoras.html), The MacTutor History of Mathematics archive, School of Mathematics and Statistics, University of St Andrews, Scotland
- *Pythagoras and the Pythagoreans, Fragments and Commentary* (http://history.hanover.edu/texts/presoc/pythagor.html), Arthur Fairbanks Hanover Historical Texts Project, Hanover College Department of History
- *Pythagoras and the Pythagoreans* (http://www.math.tamu.edu/~don.allen/history/pythag/pythag.html), Department of Mathematics, Texas A&M University
- *Pythagoras and Pythagoreanism* (http://www.newadvent.org/cathen/12587b.htm), The Catholic Encyclopedia
- *Tetraktys* (http://www.organelle.org/organelle/tetra/tetraktys.html)

- *Golden Verses of Pythagoras* (http://pythagoras.name/golden_verses_of_pythagoras.html)
- *Pythagoras on Vegetarianism* (http://www.animalrightshistory.org/animal-rights-antiquity/pythagoras.htm) Quotes from primary source historical literature on Pythagoras' view on Vegetarianism, Justice and Kindness
- *Homage to Pythagoras* (http://users.ucom.net/~vegan/)
- *Occult conception of Pythagoreanism* (http://web.archive.org/web/20091027110142/http://geocities.com/go_darkness/god-pythagorean-pentacle.html)
- *Pythagoreanism Web Article* (http://cyberspacei.com/jesusi/inlight/philosophy/western/Pythagoreanism.htm)
- *Wandering Souls: The Doctrine of Transmigration in Pythagorean Philosophy* (http://luchte.wordpress.com/wandering-souls-the-doctrine-of-transmigration-in-pythagorean-philosophy/), by Dr. James Luchte
- 45-minute documentary (http://freedocumentaries.net/media/164/Pythagoras/) about Pythagoras
- Io and Pi – theatrical play on Pythagoras' life (http://www.regolish.com/Plays.htm)
- The Symbols (http://www.sacred-texts.com/cla/gvp/gvp11.htm) of Pythagoras at The Sacred Texts online

Thales

Thales

Born	ca. 624 BC
Died	ca. 547–546 BC
School	Ionian, Milesian school, Naturalism
Main interests	Ethics, Metaphysics, Mathematics, Astronomy
Notable ideas	Water is the physis, Thales' theorem, intercept theorem

Thales of Miletus (pron.: /ˈθeɪliːz/; Greek: Θαλῆς, *Thalēs*; c. 624 BC – c. 546 BC) was a pre-Socratic Greek philosopher from Miletus in Asia Minor, and one of the Seven Sages of Greece. Many, most notably Aristotle, regard him as the first philosopher in the Greek tradition.[1] According to Bertrand Russell, "Western philosophy begins with Thales."[2] Thales attempted to explain natural phenomena without reference to mythology and was tremendously influential in this respect. Almost all of the other pre-Socratic philosophers follow him in attempting to provide an explanation of ultimate substance, change, and the existence of the world—without reference to mythology. Those philosophers were also influential, and eventually Thales' rejection of mythological explanations became an essential idea for the scientific revolution. He was also the first to define general principles and set forth hypotheses, and as a result has been dubbed the "Father of Science", though it is argued that Democritus is actually more deserving of this title.[3] [4]

In mathematics, Thales used geometry to solve problems such as calculating the height of pyramids and the distance of ships from the shore. He is credited with the first use of deductive reasoning applied to geometry, by deriving four corollaries to Thales' Theorem. As a result, he has been hailed as the first true mathematician and is the first known individual to whom a mathematical discovery has been attributed.[5]

Life

Thales was born in the city of Miletus around the mid 620s BC. Miletus was an ancient Greek Ionian city on the western coast of Asia Minor (in what is today Aydin Province of Turkey), near the mouth of the Maeander River.

Background

The dates of Thales' life are not known. However, the time of his life is roughly established by a few dateable events mentioned in the sources and an estimate of his length of life. According to Herodotus, Thales once predicted a solar eclipse which has been determined by modern methods to have been on May 28, 585 BC.[6] Diogenes Laërtius quotes the chronicle of Apollodorus of Athens as saying that Thales died at 78 in the 58th Olympiad (548–545 BC).

Diogenes Laertius states that ("according to Herodotus and Douris and Democritus") Thales' parents were Examyes and Cleobuline, both Phoenician nobles. Giving another opinion, he ultimately connects Thales' family line back to Phoenician prince Cadmus. Diogenes also reports two other stories, one that he married and had a son, Cybisthus or

Cybisthon, or adopted his nephew of the same name. The second is that he never married, telling his mother as a young man that it was too early to marry, and as an older man that it was too late. A much earlier source - Plutarch - tells the following story: Solon who visited Thales asked him the reason which kept him single. Thales answered that he did not like the idea of having to worry about children. Nevertheless, several years later Thales, anxious for family, adopted his nephew Cybisthus.[7]

Thales involved himself in many activities, taking the role of an innovator. Some say that he left no writings, others that he wrote "On the Solstice" and "On the Equinox". Neither has survived. Diogenes Laërtius quotes letters of Thales to Pherecydes and Solon, offering to review the book of the former on religion, and offering to keep company with the latter on his sojourn from Athens. Thales identifies the Milesians as Athenians.[8]

Business

Several anecdotes suggest that Thales was not solely a thinker but was also involved in business and politics. One story recounts that he bought all the olive presses in Miletus after predicting the weather and a good harvest for a particular year. In another version of the same story, Aristotle explains that Thales reserved presses ahead of time at a discount only to rent them out at a high price when demand peaked, following his predictions of a particular good harvest. Aristotle explains that Thales' objective in doing this was not to enrich himself but to prove his fellow Milesians that philosophy could be useful, contrary to what they thought.[9]

An olive mill and an olive press dating from Roman times in Capernaum, Israel.

Politics

Thales' political life had mainly to do with the involvement of the Ionians in the defense of Anatolia against the growing power of the Persians, who were then new to the region. A king had come to power in neighboring Lydia, Croesus, who was somewhat too aggressive for the size of his army. He had conquered most of the states of coastal Anatolia, including the cities of the Ionians. The story is told in Herodotus.[10]

The Lydians were at war with the Medes, a remnant of the first wave of Iranians in the region, over the issue of refuge the Lydians had given to some Scythian soldiers of fortune inimical to the Medes. The war endured for five years, but in the sixth an eclipse of the Sun (mentioned above) spontaneously halted a battle in progress (the Battle of Halys).

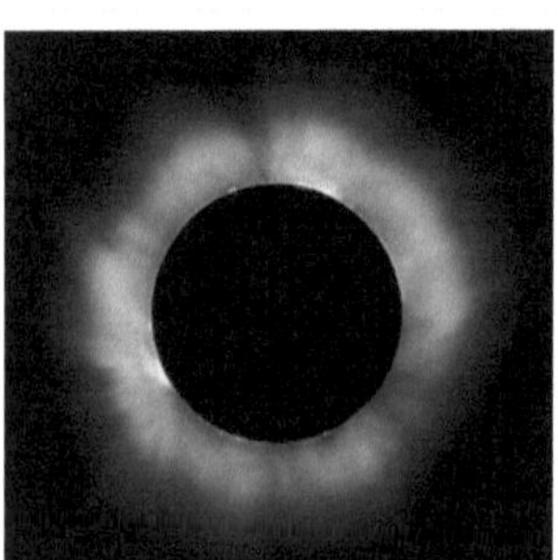

Total eclipse of the Sun

It seems that Thales had predicted this solar eclipse. The Seven Sages were most likely already in existence, as Croesus was also heavily influenced by Solon of Athens, another sage. Whether Thales was present at the battle is not known, nor are the exact terms of the prediction, but based on it the Lydians and Medes made peace immediately, swearing a blood oath.

The Medes were dependencies of the Persians under Cyrus. Croesus now sided with the Medes against the Persians and marched in the direction of Iran (with far fewer men than he needed). He was stopped by the river Halys, then unbridged. This time he had Thales with him, perhaps by invitation. Whatever his status, the king gave the problem to him, and he got the army across by digging a diversion upstream so as to reduce the flow, making it possible to ford the river. The channels ran around both sides of the camp.

The two armies engaged at Pteria in Cappadocia. As the battle was indecisive but paralyzing to both sides, Croesus marched home, dismissed his mercenaries and sent emissaries to his dependents and allies to ask them to dispatch fresh troops to Sardis. The issue became more pressing when the Persian army showed up at Sardis. Diogenes Laertius[11] tells us that Thales gained fame as a counsellor when he advised the Milesians not to engage in a symmachia, a "fighting together", with the Lydians. This has sometimes been interpreted as an alliance, but a ruler does not ally with his subjects.

Croesus was defeated before the city of Sardis by Cyrus, who subsequently spared Miletus because it had taken no action. Cyrus was so impressed by Croesus' wisdom and his connection with the sages that he spared him and took his advice on various matters.

The Ionians were now free. Herodotus says that Thales advised them to form an Ionian state; that is, a bouleuterion ("deliberative body") to be located at Teos in the center of Ionia. The Ionian cities should be demoi, or "districts". Miletus, however, received favorable terms from Cyrus. The others remained in an Ionian League of 12 cities (excluding Miletus now), and were subjugated by the Persians.

Sagacity

Diogenes Laertius[12] tells us that the Seven Sages were created in the archonship of Damasius at Athens about 582 BC and that Thales was the first sage. The same story, however, asserts that Thales emigrated to Miletus. There is also a report that he did not become a student of nature until after his political career. Much as we would like to have a date on the seven sages, we must reject these stories and the tempting date if we are to believe that Thales was a native of Miletus, predicted the eclipse, and was with Croesus in the campaign against Cyrus.

Thales had instruction from Egyptian priest. It was fairly certain that he came from a wealthy and established family, and the wealthy customarily educated their children. Moreover, the ordinary citizen, unless he was a seafaring man or a merchant, could not afford the grand tour in Egypt, and in any case did not consort with noble lawmakers such as Solon.

He did participate in some games, most likely Panhellenic, at which he won a bowl twice. He dedicated it to Apollo at Delphi. As he was not known to have been athletic, his event was probably declamation, and it may have been victory in some specific phase of this event that led to his being designated sage

Theories

The Greeks often invoked idiosyncratic explanations of natural phenomena by reference to the will of anthropomorphic gods and heroes. Thales, however, aimed to explain natural phenomena via a rational explanation that referenced natural processes themselves. For example, Thales attempted to explain earthquakes by hypothesizing that the Earth floats on water, and that earthquakes occur when the Earth is rocked by waves, rather than assuming that earthquakes were the result of supernatural processes. Thales was a Hylozoist (those who think matter is alive).[13] It is unclear whether the interpretation that he treated matter as being alive might have been mistaken for his thinking the properties of nature arise directly from material processes, more consistent with modern ideas of how properties arise as emergent characteristics of complex systems involved in the processes of evolution and developmental change.

Thales, according to Aristotle, asked what was the nature (Greek *Arche*) of the object so that it would behave in its characteristic way. Physis (φύσις) comes from phyein (φύειν), "to grow", related to our word "be".[14] *(G)natura* is the way a thing is "born",[15] again with the stamp of what it is in itself.

Aristotle[16] characterizes most of the philosophers "at first" (πρῶτον) as thinking that the "principles in the form of matter were the only principles of all things", where "principle" is arche, "matter" is hyle ("wood" or "matter", "material") and "form" is eidos.

Arche is translated as "principle", but the two words do not have precisely the same meaning. A principle of something is merely prior (related to pro-) to it either chronologically or logically. An arche (from ἄρχειν, "to rule") dominates an object in some way. If the arche is taken to be an origin, then specific causality is implied; that is, B is supposed to be characteristically B just because it comes from A, which dominates it.

The archai that Aristotle had in mind in his well-known passage on the first Greek scientists are not necessarily chronologically prior to their objects, but are constituents of it. For example, in pluralism objects are composed of earth, air, fire and water, but those elements do not disappear with the production of the object. They remain as archai within it, as do the atoms of the atomists.

What Aristotle is really saying is that the first philosophers were trying to define the substance(s) of which all material objects are composed. As a matter of fact, that is exactly what modern scientists are attempting to accomplish in nuclear physics, which is a second reason why Thales is described as the first western scientist.

Water as a first principle

Thales' most famous belief was his cosmological thesis, which held that the world started from water. Aristotle considered this belief roughly equivalent to the later ideas of Anaximenes, who held that everything in the world was composed of air.

The best explanation of Thales' view is the following passage from Aristotle's *Metaphysics*.[17] The passage contains words from the theory of matter and form that were adopted by science with quite different meanings.

> "That from which is everything that exists and from which it first becomes and into which it is rendered at last, its substance remaining under it, but transforming in qualities, that they say is the element and principle of things that are."

And again:

> "For it is necessary that there be some nature (φύσις), either one or more than one, from which become the other things of the object being saved... Thales the founder of this type of philosophy says that it is water."[18]

Aristotle's depiction of the problem of change and the definition of substance is clear. If an object changes, is it the same or different? In either case how can there be a change from one to the other? The answer is that the substance "is saved", but acquires or loses different qualities (πάθη, the things you "experience").

A deeper dip into the waters of the theory of matter and form is properly reserved to other articles. The question for this article is, how far does Aristotle reflect Thales? He was probably not far off, and Thales was probably an incipient matter-and-formist.

The essentially non-philosophic Diogenes Laertius states that Thales taught as follows:

> "Water constituted (ὑπεστήσατο, 'stood under') the principle of all things."[19]

Heraclitus Homericus[20] states that Thales drew his conclusion from seeing moist substance turn into air, slime and earth. It seems likely that Thales viewed the Earth as solidifying from the water on which it floated and which surrounded Ocean.

Beliefs in divinity

Thales applied his method to objects that changed to become other objects, such as water into earth (or so he thought). But what about the changing itself? Thales did address the topic, approaching it through lodestone and amber, which, when electrified by rubbing together, also attracts. It is noteworthy that the first particle known to carry electric charge, the electron, is named for the Greek word for amber, ἤλεκτρον (ēlektron).

How was the power to move other things without the movers changing to be explained? Thales saw a commonality with the powers of living things to act. The lodestone and the amber must be alive, and if that were so, there could be no difference between the living and the dead. When asked why he didn't die if there was no difference, he replied "because there is no difference."

Aristotle defined the soul as the principle of life, that which imbues the matter and makes it live, giving it the animation, or power to act. The idea did not originate with him, as the Greeks in general believed in the distinction between mind and matter, which was ultimately to lead to a distinction not only between body and soul but also between matter and energy.

If things were alive, they must have souls. This belief was no innovation, as the ordinary ancient populations of the Mediterranean did believe that natural actions were caused by divinities. Accordingly, the sources say that Thales believed that "all things were full of gods.".[21] In their zeal to make him the first in everything some said he was the first to hold the belief, which must have been widely known to be false.

However, Thales was looking for something more general, a universal substance of mind. That also was in the polytheism of the times. Zeus was the very personification of supreme mind, dominating all the subordinate manifestations. From Thales on, however, philosophers had a tendency to depersonify or objectify mind, as though it were the substance of animation per se and not actually a god like the other gods. The end result was a total removal of mind from substance, opening the door to a non-divine principle of action.

Classical thought, however, had proceeded only a little way along that path. Instead of referring to the person, Zeus, they talked about the great mind:

> "Thales", says Cicero,[22] "assures that *water* is the principle of all things; and that God is that Mind which shaped and created all things from water."

The universal mind appears as a Roman belief in Virgil as well:

> *"In the beginning, SPIRIT within (spiritus intus) strengthens Heaven and Earth,*
>
> *The watery fields, and the lucid globe of Luna, and then --*
>
> *Titan stars; and mind (mens) infused through the limbs*
>
> *Agitates the whole mass, and mixes itself with GREAT MATTER (magno corpore)"*[23]

Geometry

Thales was known for his innovative use of geometry. His understanding was theoretical as well as practical. For example, he said:

> Megiston topos: hapanta gar chorei (Μέγιστον τόπος· άπαντα γαρ χωρεί)
>
> "Space is the greatest thing, as it contains all things"

Topos is in Newtonian-style space, since the verb, chorei, has the connotation of yielding before things, or spreading out to make room for them, which is extension. Within this extension, things have a position. Points, lines, planes and solids related by distances and angles follow from this presumption.

Thales understood similar triangles and right triangles, and what is more, used that knowledge in practical ways. The story is told in DL (loc. cit.) that he measured the height of the pyramids by their shadows at the moment when his own shadow was equal to his height. A right triangle with two equal legs is a 45-degree right triangle, all of which are similar. The length of the pyramid's shadow measured from the center of the pyramid at that moment must have been equal to its height.

This story indicates that he was familiar with the Egyptian seked, or seqed - the ratio of the run to the rise of a slope (cotangent). The seked is at the base of problems 56, 57, 58, 59 and 60 of the Rhind papyrus - an ancient Egyptian mathematics document.

In present day trigonometry, cotangents require the same units for run and rise (base and perpendicular), but the papyrus uses cubits for rise and palms for run, resulting in different (but still characteristic) numbers. Since there were 7 palms in a cubit, the seked was 7 times the cotangent.

To use an example often quoted in modern reference works, suppose the base of a pyramid is 140 cubits and the angle of rise 5.25 seked. The Egyptians expressed their fractions as the sum of fractions, but the decimals are sufficient for the example. What is the rise in cubits? The run is 70 cubits, 490 palms. X, the rise, is 490 divided by 5.25 or $93\frac{1}{3}$ cubits. These figures sufficed for the Egyptians and Thales. We would go on to calculate the cotangent as 70 divided by $93\frac{1}{3}$ to get 3/4 or .75 and looking that up in a table of cotangents find that the angle of rise is a few minutes over 53 degrees.

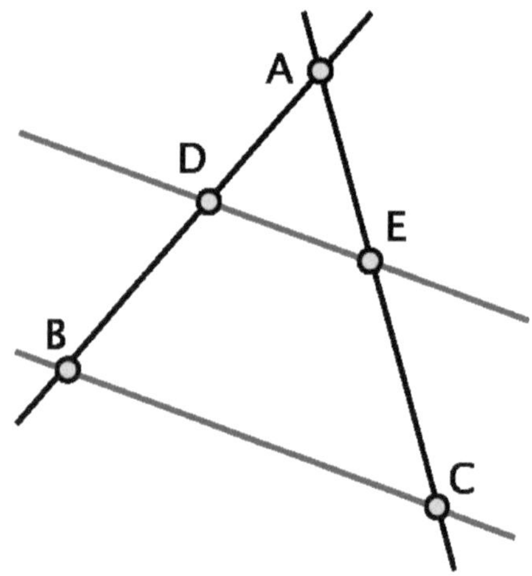

Thales' Theorem:

Whether the ability to use the seked, which preceded Thales by about 1000 years, means that he was the first to define trigonometry is a matter of opinion. More practically Thales used the same method to measure the distances of ships at sea, said Eudemus as reported by Proclus ("in Euclidem"). According to Kirk & Raven (reference cited below), all you need for this feat is three straight sticks pinned at one end and knowledge of your altitude. One stick goes vertically into the ground. A second is made level. With the third you sight the ship and calculate the seked from the height of the stick and its distance from the point of insertion to the line of sight.

The seked is a measure of the angle. Knowledge of two angles (the seked and a right angle) and an enclosed leg (the altitude) allows you to determine by similar triangles the second leg, which is the distance. Thales probably had his own equipment rigged and recorded his own seked, but that is only a guess.

Thales' Theorem is stated in another article. (Actually there are two theorems called Theorem of Thales, one having to do with a triangle inscribed in a circle and having the circle's diameter as one leg, the other theorem being also called the intercept theorem.) In addition Eudemus attributed to him the discovery that a circle is bisected by its diameter, that the base angles of an isosceles triangle are equal and that vertical angles are equal. According to a historical Note,[24] when Thales visited Egypt, he observed that whenever the Egyptians drew two intersecting lines, they would measure the vertical angles to make sure that they were equal. Thales concluded that one could prove that all vertical angles are equal if one accepted some general notions such as: all straight angles are equal, equals added to equals are equal, and equals subtracted from equals are equal. It would be hard to imagine civilization without these theorems.

Interpretations

In the long sojourn of philosophy there has existed hardly a philosopher or historian of philosophy who did not mention Thales and try to characterize him in some way. He is generally recognized as having brought something new to human thought. Mathematics, astronomy and medicine already existed. Thales added something to these different collections of knowledge to produce a universality, which, as far as writing tells us, was not in tradition before, but resulted in a new field.

Ever since, interested persons have been asking what that new something is. Answers fall into (at least) two categories, the theory and the method. Once an answer has been arrived at, the next logical step is to ask how Thales compares to other philosophers, which leads to his classification (rightly or wrongly).

Theory

The most natural epithets of Thales are "materialist" and "naturalist", which are based on ousia and physis. The Catholic Encyclopedia notes that Aristotle called him a physiologist, with the meaning "student of nature."[25] On the other hand, he would have qualified as an early physicist, as did Aristotle. They studied corpora, "bodies", the medieval descendants of substances.

Most agree that Thales' stamp on thought is the unity of substance, hence Bertrand Russell:[26]

> "The view that all matter is one is quite a reputable scientific hypothesis."
>
> "...But it is still a handsome feat to have discovered that a substance remains the same in different states of aggregation."

Russell was only reflecting an established tradition; for example: Nietzsche, in his *Philosophy in the Tragic Age of the Greeks*, wrote:[27]

> "Greek philosophy seems to begin with an absurd notion, with the proposition that *water* is the primal origin and the womb of all things. Is it really necessary for us to take serious notice of this proposition? It is, and for three reasons. First, because it tells us something about the primal origin of all things; second, because it does so in language devoid of image or fable, and finally, because contained in it, if only embryonically, is the thought, 'all things are one.'"

This sort of materialism, however, should not be confused with deterministic materialism. Thales was only trying to explain the unity observed in the free play of the qualities. The arrival of uncertainty in the modern world made possible a return to Thales; for example, John Elof Boodin writes ("God and Creation"):

> "We cannot read the universe from the past..."

Boodin defines an "emergent" materialism, in which the objects of sense emerge uncertainly from the substrate. Thales is the innovator of this sort of materialism.

The Rise of Theoretical Inquiry

In the West, Thales represents a new kind of inquiring community as well. Edmund Husserl[28] attempts to capture the new movement as follows. Philosophical man is a "new cultural configuration" based in stepping back from "pregiven tradition" and taking up a rational "inquiry into what is true in itself;" that is, an ideal of truth. It begins with isolated individuals such as Thales, but they are supported and cooperated with as time goes on. Finally the ideal transforms the norms of society, leaping across national borders.

Classification

The term "Pre-Socratic" derives ultimately from the philosopher Aristotle, who distinguished the early philosophers as concerning themselves with substance.

Diogenes Laertius on the other hand took a strictly geographic and ethnic approach. Philosophers were either Ionian or Italian. He used "Ionian" in a broader sense, including also the Athenian academics, who were not Pre-Socratics. From a philosophic point of view, any grouping at all would have been just as effective. There is no basis for an Ionian or Italian unity. Some scholars, however, concede to Diogenes' scheme as far as referring to an "Ionian" school. There was no such school in any sense.

The most popular approach refers to a Milesian school, which is more justifiable socially and philosophically. They sought for the substance of phenomena and may have studied with each other. Some ancient writers qualify them as Milesioi, "of Miletus."

Influence on others

Thales (Electricity), sculpture from "The Progress of Railroading" (1908), main facade of Union Station (Washington, DC)

Thales had a profound influence on other Greek thinkers and therefore on Western history. Some believe Anaximander was a pupil of Thales. Early sources report that one of Anaximander's more famous pupils, Pythagoras, visited Thales as a young man, and that Thales advised him to travel to Egypt to further his philosophical and mathematical studies.

Many philosophers followed Thales' lead in searching for explanations in nature rather than in the supernatural; others returned to supernatural explanations, but couched them in the language of philosophy rather than of myth or of religion.

Looking specifically at Thales' influence during the pre-Socratic era, it is clear that he stood out as one of the first thinkers who thought more in the way of *logos* than *mythos*. The difference between these two more profound ways of seeing the world is that *mythos* is concentrated around the stories of holy origin, while *logos* is concentrated around the argumentation. When the mythical man wants to explain the world the way he sees it, he explains it based on gods and powers. Mythical thought does not differentiate between things and persons and furthermore it does not differentiate between nature and culture. The way a *logos* thinker would present a world view is radically different from the way of the mythical thinker. In its concrete form, *logos* is a way of thinking not only about individualism, but also the abstract. Furthermore, it focuses on sensible and continuous argumentation. This lays the foundation of philosophy and its way of explaining the world in terms of abstract argumentation, and not in the way of gods and mythical stories.

Sources

Our sources on the Milesian philosophers (Thales, Anaximander, and Anaximenes) were either roughly contemporaneous (such as Herodotus) or lived within a few hundred years of his passing. Moreover, they were writing from a tradition that was well-known. Most modern dissension comes from trying to interpret what we know, in particular, distinguishing legend from fact.

The main secondary source concerning the details of Thales' life and career is Diogenes Laertius, "*Lives of Eminent Philosophers*".[29] This is primarily a biographical work, as the name indicates. Compared to Aristotle, Diogenes is not much of a philosopher. He is the one who, in the Prologue to that work, is responsible for the division of the early philosophers into "Ionian" and "Italian", but he places the Academics in the Ionian school and otherwise evidences considerable disarray and contradiction, especially in the long section on forerunners of the "Ionian School". Diogenes quotes two letters attributed to Thales, but Diogenes wrote some eight centuries after Thales' death and that his sources often contained "unreliable or even fabricated information",[30] hence the concern for separating fact from legend in accounts of Thales.

Most philosophic analyses of the philosophy of Thales come from Aristotle, a professional philosopher, tutor of Alexander the Great, who wrote 200 years after Thales death. Aristotle, judging from his surviving books, does not seem to have access to any works by Thales, although he probably had access to works of other authors about Thales, such as Herodotus, Hecataeus, Plato etc., as well as others whose work is now extinct. It was Aristotle's express goal to present Thales work not because it was significant in itself, but as a prelude to his own work in natural philosophy.[31] Geoffrey Kirk and John Raven, English compilers of the fragments of the Pre-Socratics, assert that Aristotle's "judgments are often distorted by his view of earlier philosophy as a stumbling progress toward the truth that Aristotle himself revealed in his physical doctrines."[32] There was also an extensive oral tradition. Both the oral and the written were commonly read or known by all educated men in the region.

Aristotle's philosophy had a distinct stamp: it professed the theory of matter and form, which modern scholastics have dubbed hylomorphism. Though once very widespread, it was not generally adopted by rationalist and modern science, as it mainly is useful in metaphysical analyses, but does not lend itself to the detail that is of interest to modern science. It is not clear that the theory of matter and form existed as early as Thales, and if it did, whether Thales espoused it.

See also

- Know thyself
- Material monism
- The Astrologer who Fell into a Well

Notes

[1] Aristotle, Metaphysics Alpha, 983b18.

[2] Russell, Bertrand (1945). *The History of Western Philosophy*. New York: Simon and Schuster.

[3] Singer, C. (2008). *A Short History of Science to the 19th century*. Streeter Press. p. 35.

[4] Needham, C. W. (1978). *Cerebral Logic: Solving the Problem of Mind and Brain*. Loose Leaf. p. 75. ISBN 0-398-03754-X.

[5] (Boyer 1991, "Ionia and the Pythagoreans" p. 43)

[6] Herodotus, 1.74.2 (http://www.perseus.tufts.edu/cgi-bin/ptext?lookup=Hdt.+1.74.1), and A. D. Godley's footnote 1; Pliny, 2.9 (12) (http://www.perseus.tufts.edu/cgi-bin/ptext?lookup=Plin.+Nat.+2.9) and Bostock's footnote 2.

[7] Plutarch (1952). "Solon" (http://www.amazon.com/Great-Books-Western-World-Vol/dp/B000L3QMIK). In Robert Maynard Hutchins. *Lives*. Great Books of the Western World. **14**. Chicago: William Benton. pp. 66. .

[8] Diogenes Laërtius, 1.43, 44.

[9] Aristotle, Politics 1259a (http://www.perseus.tufts.edu/cgi-bin/ptext?lookup=Aristot.+Pol.+1.1259a)

[10] Book 1

[11] 1.25

[12] 1.22

[13] Farrington, B., 1944 Greek Science. Pelican

[14] English physics (http://www.bartleby.com/61/77/P0277700.html) comes from it, but the latter is a Greek loan. In addition the quite ancient native English word be (http://www.bartleby.com/61/roots/IE62.html) comes from the same Indo-European root.

[15] The initial g of the archaic Latin gives the root away as *genə- (http://www.bartleby.com/61/roots/IE143.html), "beget."

[16] Metaphysics 983b6

[17] 983 b6 8-11

[18] Lines 17-21 with gaps.

[19] Work cited, paragraph 27.

[20] Quaes. Hom. 22, not the same as Heraclitus of Ephesus

[21] Aristotle, *De Anima*, 411a7. For other ancient sources see the discussion in Kirk and Raven, *The Presocratic Philosophers*, 93-7.

[22] *De natura Deorum*, i.,10

[23] Virgil:"Aeneid," vi., 724-727.

[24] William G. Shute, William W. Shirk, George F. Porter, *Plane and Solid Geometry*, American Book Company (1960) pp. 25-27

[25] Turner, *Catholic Encyclopedia.*

[26] *Wisdom of the West*

[27] § 3

[28] *The Vienna Lecture*

[29] Translation of his biography on Thales: Thales (http://www.classicpersuasion.org/pw/diogenes/dlthales.htm), classicpersuasion site; original Greek text, under ΘΑΛΗΣ (http://www.mikrosapoplous.gr/dl/dl01.html#thales), the Library of Ancient Texts Online site.

[30] See McKirahan, Richard D., Jr. (1994). *Philosophy Before Socrates*. Indianapolis: Hackett. p. 5. ISBN 0-87220-176-7.

[31] See Aristotle, Metaphysics Alpha, 983b 1-27.

[32] Kirk and Raven, The Presocratic Philosophers, Second Edition (Cambridge University Press, 1983) 3.

References

- Burnet, John (1957) [1892]. *Early Greek Philosophy*. The Meridian Library. (reprinted from the 4th edition, 1930; the first edition was published in 1892). An online presentation of the Third Edition (http://faculty.evansville.edu/tb2/courses/phil211/burnet/) can be found in the Online Books Library of the University of Pennsylvania.
- Diogenes Laërtius, *Life of Thales*, translated by Robert Drew Hicks (1925).
- Herodotus; *Histories*, A. D. Godley (translator), Cambridge: Harvard University Press, 1920; ISBN 0-674-99133-8. Online version (http://www.perseus.tufts.edu/cgi-bin/ptext?lookup=Hdt.+1.1.0) at the Perseus Digital Library.
- Hans Joachim Störig: *Kleine Weltgeschichte der Philosophie*. Fischer, Frankfurt/M. 2004, ISBN 3-596-50832-0.
- Kirk, G.S.; Raven, J.E. (1957). *The Presocratic Philosophers*. Cambridge: University Press. (subsequently reprinted)
- G. E. R. Lloyd. *Early Greek Science: Thales to Aristotle.*
- Nahm, Milton C. (1962) [1934]. *Selections from Early Greek Philosophy*. Appleton-Century-Crofts, Inc.
- Pliny the Elder; *The Natural History* (eds. John Bostock, M.D., F.R.S. H.T. Riley, Esq., B.A.) London. Taylor and Francis, Red Lion Court, Fleet Street. (1855). Online version (http://www.perseus.tufts.edu/cgi-bin/ptext?lookup=Plin.+Nat.+toc) at the Perseus Digital Library.
- Turner, William. "Ionian School of Philosophy." *The Catholic Encyclopedia*. Vol. 8. New York: Robert Appleton Company, 1910. <http://www.newadvent.org/cathen/08092a.htm>.

Further reading

- Morse, Sidney H.; Marvin, Joseph B., *The Radical* (http://books.google.com/books?id=kIAQAAAAIAAJ&pg=PA170&lpg=PA170&dq=thales+sound+mind+in+sound+body&source=bl&ots=73Il9ErYW3&sig=Bdpml5rYncGWPcq8y1AqYoSRy7c#PPA170,M1), A. Williams & Co., 1868. Cf. p. 170 and on.

External links

- Thales of Miletus from The Internet Encyclopedia of Philosophy (http://www.iep.utm.edu/t/thales.htm)
- Thales of Miletus (http://www-groups.dcs.st-and.ac.uk/~history/Mathematicians/Thales.html) from the MacTutor History of Mathematics archive
- Livius (http://www.livius.org), Thales of Miletus (http://www.livius.org/th/thales/thales.html) by Jona Lendering
- Thales (http://www.philosophy.gr/presocratics/thales.htm)
- Thales' Theorem - Math Open Reference (http://www.mathopenref.com/thalestheorem.html) With interactive animation
- Thales biography by Charlene Douglass (http://www.mathopenref.com/thales.html) With extensive bibliography.

Hypatia

The Martyrdom of Hypatia - Charles William Mitchell

Born	ca. AD 351–370Alexandria
Died	AD 415Alexandria
Era	Ancient philosophy
Region	Alexandria, Egypt
School	Neoplatonism
Main interests	Mathematics, astronomy

Hypatia (ca. AD 350–370–March 415) (pron.: /haɪˈpeɪʃə/ *hy-**PAY**-shə*; Ancient Greek: Ὑπατία; *Hypatía*) was a Greek Neoplatonist philosopher in Roman Egypt who was the first historically noted woman in mathematics.[1] As head of the Platonist school at Alexandria, she also taught philosophy and astronomy.[2] [3] [4] [5] [6]

As a Neoplatonist philosopher, she belonged to the mathematic tradition of the Academy of Athens, as represented by Eudoxus of Cnidus;[7] she was of the intellectual school of the 3rd century thinker Plotinus, which encouraged logic and mathematical study in place of empirical enquiry and strongly encouraged law in place of nature.[1]

According to the only contemporary source, Hypatia was murdered by a Christian mob after being accused of exacerbating a conflict between two prominent figures in Alexandria: the governor Orestes and the Bishop of Alexandria.[8] Kathleen Wider proposes that the murder of Hypatia marked the end of Classical antiquity,[9] and Stephen Greenblatt observes that her murder "effectively marked the downfall of Alexandrian intellectual life".[10] On the other hand, Maria Dzielska and Christian Wildberg note that Hellenistic philosophy continued to flourish in the 5th and 6th centuries, and perhaps until the age of Justinian.[11]

Life

The mathematician and philosopher Hypatia of Alexandria was the daughter of the mathematician Theon Alexandricus (ca. 335–405).[12] She was educated at Athens and in Italy. Around AD 400, she became head of the Platonist school at Alexandria,[13] [14] [15] where she imparted the knowledge of Plato and Aristotle to any student; the pupils included pagans, Christians, and foreigners.[1] [16] [17]

The contemporary 5th-century sources do identify Hypatia of Alexandria as a practitioner and teacher of the philosophy of Plato and Plotinus, but, two hundred years later, the 7th-century Egyptian Coptic bishop John of Nikiû identified her as a Hellenistic pagan and that "she was devoted at all times to magic, astrolabes and instruments of music, and she beguiled many people through her Satanic wiles".[18] [19] Not all Christians were as hostile towards her as John of Nikiu or the monks who killed her: some Christians even used Hypatia as symbolic of Virtue.[1]

The Byzantine *Suda* encyclopaedia reported that Hypatia was "the wife of Isidore the Philosopher" (apparently Isidore of Alexandria);[16] however, Isidore of Alexandria was not born until long after Hypatia's death, and no other philosopher of that name contemporary with Hypatia is known.[20] The *Suda* also stated that "she remained a virgin" and that she rejected a suitor with her menstrual rags, saying that they demonstrated "nothing beautiful" about carnal desire, an example of a Christian source using Hypatia as a symbol of Virtue.[16] [21] [22]

Hypatia corresponded with former pupil Synesius of Cyrene, who was tutored by her in the philosophical school of Platonism and later became bishop of Ptolemais in AD 410, an exponent of the Christian Holy Trinity doctrine.[23] Together with the references by the pagan philosopher Damascius, these are the extant records left by Hypatia's pupils at the Platonist school of Alexandria.[24] The contemporary Christian historiographer Socrates Scholasticus described her in *Ecclesiastical History*:

> There was a woman at Alexandria named Hypatia, daughter of the philosopher Theon, who made such attainments in literature and science, as to far surpass all the philosophers of her own time. Having succeeded to the school of Plato and Plotinus, she explained the principles of philosophy to her auditors, many of whom came from a distance to receive her instructions. On account of the self-possession and ease of manner which she had acquired in consequence of the cultivation of her mind, she not infrequently appeared in public in the presence of the magistrates. Neither did she feel abashed in going to an assembly of men. For all men on account of her extraordinary dignity and virtue admired her the more.[1]
>
> —Socrates Scholasticus, Ecclesiastical History

Death

Events leading to her murder

Two widely cited, but divergent texts describe the feud between Orestes, the prefect (or Governor) of Alexandria and Cyril, the Bishop of Alexandria. The feud and the city-wide anger it provoked ultimately brought about the death of Hypatia.

One source, the *Historia Ecclesiastica* (or "Ecclesiastical History") was written by Socrates Scholasticus (who was himself a Christian), some time shortly after Hypatia's death in AD 415. Scholasticus gives a more complete, less biased account of the feud between Orestes and Cyril, and the role Hypatia played in the feud that resulted in her death.

The other source, *The Chronicle*,[25] written by John of Nikiu in Egypt, around 650 AD, demonizes Hypatia and Orestes directly, while validating all Christians involved in the events Nikiu describes. *The Chronicle*, in being more biased on the matter of the historical feud, omits several points of the narrative that are detailed in Scholasticus's account.

***Ecclesiastical History*, Socrates Scholasticus**[26]

Orestes, the governor of Alexandria, and Cyril, the Bishop of Alexandria, found themselves in a bitter feud in which Hypatia would come to be one of the main points of contention. The feud, which took place in 415 AD, began over the matter of Jewish dancing exhibitions in Alexandria. Since these exhibitions attracted large crowds and were commonly prone to civil disorder of varying degrees, Orestes published an edict which outlined new regulations for such gatherings and posted it in the city's theater. Soon after, crowds gathered to read the edict, angry over the new regulations that had been imposed upon them. At one such gathering, Hierax, a Christian and devout follower of Cyril, read the edict and applauded the new regulations, which many people felt was an attempt to incite the crowd into sedition. In what Scholasticus suspected as Orestes' "...jealousy [of] the growing power of the bishops...[which] encroached on the jurisdiction of the authorities...", Orestes immediately ordered Hierax to be seized and publicly tortured in the theater.

Upon hearing of this, Cyril threatened the Jews of Alexandria with "the utmost severities" if harassment of Christians was not ceased at once. In response, the Jews of Alexandria grew only more furious over Cyril's threat, and in their anger they eventually resorted to violence against the Christians. They plotted to flush the Christians out at night by running through the streets, claiming that the Church of Alexander was on fire. When the Christians responded to what they were led to believe was the burning down of their church, "the Jews immediately fell upon and slew them", using rings to recognize one another in the dark, while killing everyone else in sight. When the morning came, the Jews of Alexandria could not hide their guilt, and Cyril, along with many of his followers, took to the city's synagogues in search of the perpetrators of the night's massacre.

After Cyril found all of the Jews in Alexandria, he ordered them to be stripped of all their possessions, banished them from Alexandria, and allowed the remaining citizens to pillage the goods they left behind. With Cyril's banishment of the Jews, "Orestes [...] was filled with great indignation at these transactions, and was excessively grieved that a city of such magnitude should have been suddenly bereft of so large a portion of its population...". Because of this, the feud between Cyril and Orestes only grew stronger, and both men wrote to the emperor regarding the situation. Eventually, Cyril attempted to reach out to Orestes through several peace overtures, including attempted mediation and, when that failed, showed him the Gospels. Nevertheless, Orestes remained unmoved by such gestures.

Meanwhile, approximately 500 monks, who resided in the mountains of Nitria, and were "of a very fiery disposition", heard of the ongoing feud between the Governor and Bishop, and shortly thereafter descended into Alexandria, armed and prepared to fight alongside Cyril. Upon their arrival in Alexandria, the monks quickly intercepted Orestes' chariot in town and proceeded to bombard and harass him, calling him a pagan idolater. In response to such allegations, Orestes countered that he was actually a Christian, and had even been baptized by Atticus, the Bishop of Constantinople. The monks paid little attention to Orestes' claims of Christianity, and one of the monks, by the name of Ammonius, struck Orestes in the head with a rock, which caused him to bleed profusely. At this point, Orestes' guards fled for fear of their lives, but a nearby crowd of Alexandrians came to his aid, and Ammonius was subsequently secured and ordered to be tortured for his actions. Upon excessive torture, Ammonius died. Following the death of Ammonius, Cyril ordered that he henceforth be remembered as a martyr. Such a proclamation did not sit well with "sober-minded" Christians, as Scholasticus pointed out, seeing that he "suffered the punishment due to his rashness...[not because] he would not deny Christ", and this fact, according to Scholasticus, became more apparent to Cyril through general lack of enthusiasm for Ammonius's case for martyrdom.

Scholasticus then introduces Hypatia, the female philosopher of Alexandria and woman who would become a target of the Christian anger that grew over the feud. Daughter of Theon, and a teacher trained in the philosophical schools of Plato and Plotinus, she was admired by most men for her dignity and virtue. Of the anger she provoked among Christians, Scholasticus writes, Hypatia ultimately fell "victim to the political jealousy which at the time prevailed" - Orestes was known to seek her counsel, and a rumor spread among the Christian community of Alexandria in which she was blamed for his unwillingness to reconcile with Cyril. Therefore, a mob of Christians gathered, led by a reader (i.e. a minor cleric) named Peter whom Scholasticus calls a fanatic. They kidnapped Hypatia on her way home and took her to the "Church called Caesareum. They then completely stripped her, and then murdered her with tiles". Socrates Scholasticus was hence interpreted as saying that, while she was still alive, Hypatia's flesh was torn off using oyster shells (tiles; the Greek word is *ostrakois*, which literally means "oystershells" but the word was also used for brick tiles on the roofs of houses). Afterward, the men proceeded to mutilate her, and finally burn her limbs. When news broke of Hypatia's murder, it provoked great public denouncement; not only against Cyril, but against the whole Alexandrian Christian community. Scholasticus closes with a lament: "Surely nothing can be farther from the spirit of Christianity than the allowance of massacres, fights, and transactions of that sort".

Chronicle, John of Nikiu[27]

Bishop John of Nikiu, who lived several hundred years after the events he describes, writes bitterly of Hypatia, claiming that "she beguiled many people through (her) Satanic wiles". Orestes, who Nikiu writes, was himself a victim of Hypatia's demonic charm, regularly honored her, and took to abandoning the Christian Church in order to follow her teachings more closely. Moreover, Orestes himself persuaded others to leave the Church in favor of Hypatia's philosophical teachings, and went as far as to host such "unbelievers" at his house.

One day, Orestes published an edict "regarding public exhibitions in the city of Alexandria", and all citizens gathered to read Orestes's edict. Cyril, curious to see why the edict caused such an uproar, sent Hierax, a "Christian possessing understanding and intelligence", who although opposed to paganism, did as Cyril asked and went to learn the nature of Orestes's edict. Meanwhile, the Jews who gathered in anger over the edict, believed that Hierax had only come for the sake of provocation (which, according to Scholasticus's text, was Hierax's intent). Upon this assumption, Orestes had Hierax punished for a crime for which "he was wholly guiltless".

For the punishment and torture of Hierax, as well as the death of several monks, including Ammonius, Cyril grew increasingly furious with Orestes. (Here, Nikiu blatantly ignores the assault of Orestes by the 500 monks, of which Ammonius played an active role in bringing about his torture and death.) Cyril then warned the Jews against any further harm upon the Christians. However, with the support of Orestes (which was in no way implied by Scholasticus), the Jews felt confident in defying Cyril's authority, and so one night ran through the streets proclaiming: "The church of the apostolic Athanasius (Alexander) is on fire: come to its succour, all ye Christians". The Christians responded to the claims only to be slaughtered by the Jews in a coordinated ambush.

The next morning, all remaining Christians of the town came to Cyril with news of the massacre, after which Cyril marched with them to purge the Jews from Alexandria. In so doing, Cyril allowed the pillaging of their possessions, and soon after purified all the synagogues in the city and made them into Churches (Scholasticus makes no mention of "purifying" the Synagogues). In the expulsion of the Jews, Orestes was unable to offer them any assistance.

Shortly thereafter, a group of Christians, under Peter the magistrate, went looking for Hypatia, the "pagan woman who had beguiled the people of the city and the prefect through her enchantments". They found her sitting in a chair, at which point they seized and brought her to "the great church, named Caesarion", where they proceeded to rip the clothes off of her body. Following this, they took to dragging her through the streets of Alexandria until she died. Once she had died, they burned her remains. Nikiu's description of Hypatia's death also differs from Scholasticus's interpretation. Following the death of Hypatia, Bishop Cyril was named "the new Theophilus". With the death of Hypatia, Nikiu writes, the Christians had expelled the last remnant of pagan idolatry.

Socrates Scholasticus (born after 380 AD, died after 439 AD))	**John of Nikiû (7th century)**
Yet even she fell a victim to the political jealousy which at that time prevailed. For as she had frequent interviews with Orestes, it was calumniously reported among the Christian populace, that it was she who prevented Orestes from being reconciled to the bishop. Some of them therefore, hurried away by a fierce and bigoted zeal, whose ringleader was a reader named Peter, waylaid her returning home, and dragging her from her carriage, they took her to the church called Caesareum, where they completely stripped her, and then murdered her with tiles. After tearing her body in pieces, they took her mangled limbs to a place called Cinaron, and there burnt them.[28]	And, in those days, there appeared in Alexandria a female philosopher, a pagan named Hypatia, and she was devoted at all times to magic, astrolabes, and instruments of music, and she beguiled many people through Satanic wiles . . . A multitude of believers in God arose under the guidance of Peter the Magistrate . . . and they proceeded to seek for the pagan woman who had beguiled the people of the city and the Prefect through her enchantments. And when they learnt the place where she was, they proceeded to her and found her . . . they dragged her along till they brought her to the great church, named Caesareum. Now this was in the days of the fast. And they tore off her clothing and dragged her . . . through the streets of the city till she died. And they carried her to a place named Cinaron, and they burned her body with fire.[18]

Works

No written work, widely recognized by scholars as Hypatia's own, has survived to the present time. Many of the works commonly attributed to her are believed to have been collaborative works with her father, Theon Alexandricus, this kind of authorial uncertainty being typical for female philosophers in Antiquity.[29]

A partial list of Hypatia's works as mentioned by other antique and medieval authors or as posited by modern authors:

- A commentary on the 13-volume *Arithmetica* by Diophantus.[16]
- A commentary on the *Conics* of Apollonius.[16]
- Edited the existing version of Ptolemy's *Almagest*.[30]
- Edited her father's commentary on Euclid's *Elements*[31]
- She wrote a text "*The Astronomical Canon*".[16] (Either a new edition of Ptolemy's *Handy Tables* or the aforementioned *Almagest*.)[32] [33]

Her contributions to science are reputed to include the charting of celestial bodies[6] and the invention of the hydrometer,[34] used to determine the relative density (or specific gravity) of liquids. However, the hydrometer was invented before Hypatia, and already known in her time.[35] [36]

Her student Synesius, bishop of Cyrene, wrote a letter describing his construction of an astrolabe.[37] Earlier astrolabes predate that of Synesius by at least a century,[38] [39] and Hypatia's father had gained fame for his treatise on the subject.[40] However, Synesius claimed that his was an improved model.[41] Synesius also sent Hypatia a letter describing a hydrometer, and requesting her to have one constructed for him.[42]

Legacy

Late Antiquity to the Age of Reason

Shortly after her murder, there appeared under Hypatia's name a forged anti-Christian letter.[43] The Neoplatonist historian Damascius (ca. AD 458–538) was "anxious to exploit the scandal of Hypatia's death", and attributed responsibility for her murder to Bishop Cyril and his Christian followers; that historical account is contained in the *Suda*.[44] Damascius's account of the Christian murder of Hypatia is the sole historical source attributing responsibility to Bishop Cyril.[45] Maria Dzielska proposes that the bishop's body guards might have murdered Hypatia.[46]

The intellectual Eudokia Makrembolitissa (1021–1096), the second wife of Byzantine Emperor Constantine X Doukas, was described by the historian Nicephorus Gregoras as a "second Hypatia".[47]

Centuries later, the early 18th-century deist scholar John Toland used the murder of Hypatia as the basis for the anti-Catholic tract *Hypatia: Or the History of a most beautiful, most vertuous, most learned, and every way accomplish'd Lady; who was torn to pieces by the Clergy of Alexandria, to gratify the pride, emulation, and cruelty of their Archbishop, commonly, but undeservedly, stil'd St. Cyril*.[48]

In turn, the Christians defended themselves from Toland with *The History of Hypatia, a most Impudent School-Mistress of Alexandria: Murder'd and torn to Pieces by the Populace, in Defence of Saint Cyril and the Alexandrian Clergy from the Aspersions of Mr. Toland*, by Thomas Lewis, in 1721.[49]

19th century

In the 19th century, interest in the "literary legend of Hypatia" began to rise.[50] Diodata Saluzzo Roero's 1827 *Ipazia ovvero delle Filosofie* suggested that Cyril had actually converted Hypatia to Christianity, and that she had been killed by a "treacherous" priest.

In 1843, German authors Soldan and Heppe argued in their highly influential *History of the Witchcraft Trials* that Hypatia may have been, in effect, the first famous "witch" punished under Christian authority (see Witch-hunt).[51]

In his 1847 *Hypatie* [52] and 1857 *Hypatie et Cyrille* [53], French poet Charles-Marie-René Leconte de Lisle portrayed Hypatia as the epitome of "vulnerable truth and beauty".[54]

Charles Kingsley's 1853 novel *Hypatia – or New Foes with an Old Face*, which portrayed the scholar as a "helpless, pretentious, and erotic heroine",[55] recounted her conversion by a Jewish-Christian named Raphael Aben-Ezra after supposedly becoming disillusioned with Orestes.

In 1867, the early photographer Julia Margaret Cameron created a portrait of the scholar as a young woman.[56]

20th century

Some authors mention her in passing, such as Marcel Proust, who dropped her name in the last sentence of "Madame Swann at Home," the first section of *Within a Budding Grove*.

Some characters are named after her, such as Hypatia Cade, a precocious child and main character in the science fiction novel *The Ship Who Searched* by Mercedes Lackey and Anne McCaffrey.

Rinne Groff's 2000 play *The Five Hysterical Girls Theorem* features a character named Hypatia who lives silently, in fear that she will suffer the fate of her namesake.

Hypatia is the name of a 'shipmind' (ship computer) in *The Boy Who Would Live Forever*, a novel in Frederik Pohl's Heechee series.

Umberto Eco's novel *Baudolino* sees the protagonist meet a secluded society of satyr-like creatures who all take their name and philosophy from Hypatia.

A fictional version of the historic character appears in several works and indeed series, such as

- The *Heirs of Alexandria series* written by Mercedes Lackey, Eric Flint and Dave Freer, is an alternate history in which Hypatia was converted to Christianity by John Chrysostom, which saved her life and enabled her to stop the mob from destroying the Library of Alexandria; the books, taking place in the world of 1530 resulting from the above, include copies from the alternate Hypatia's influential correspondence with John and St. Augustine.
- The Corto Maltese adventure *Fable of Venice*, by characteristic superposition of anachronistic elements, sees Hypatia preside over an intellectual salon in pre-Fascist Italy;
- As a recurring character in Mark London Williams' juvenile fiction *Danger Boy*.

She also appears, briefly, as one of the kidnapped scientists and philosophers in the Doctor Who episode Time and the Rani.

American astronomer Carl Sagan, in *Cosmos: A Personal Voyage*, gave a detailed speculative description of Hypatia's death, linking it with the destruction of the Library of Alexandria.

A more scholarly historical study of her, *Hypatia of Alexandria* by Maria Dzielska (translated into English by F. Lyra, published by Harvard University Press), was named by *Choice Magazine* as an "Outstanding Academic Book of 1995, Philosophy Category".

She has been claimed by second wave feminism, most prominently as *Hypatia: A Journal of Feminist Philosophy*, published since 1986 by Indiana University Press.

Judy Chicago's large-scale *The Dinner Party* awards her a place-setting, and other artistic works draw on or are based on Hypatia.

A central character in Iain Pears' *The Dream of Scipio* is a woman philosopher clearly modeled on (though not identical with) Hypatia.

The last two centuries have seen Hypatia's name honored in the sciences, especially astronomy. 238 Hypatia, a main belt asteroid discovered in 1884, was named for her. The lunar crater *Hypatia* was named for her, in addition to craters named for her father Theon and for Cyril. The 180 km *Rimae Hypatia* is located north of the crater, one degree south of the equator, along the Mare Tranquillitatis.[57]

By the end of the 20th century Hypatia's name was applied to projects ranging in scope from an Adobe typeface (Hypatia Sans Pro),[58] to a cooperative community house in Madison, Wisconsin. A genus of moth also bears her name.

21st century

Her life continues to be fictionalized by authors in many countries and languages. Two recent examples are *Ipazia, scienziata alessandrina* by Adriano Petta (translated from the Italian in 2004 as *Hypatia: Scientist of Alexandria*), and *Hypatia y la eternidad* (Hypatia and Eternity) by Ramon Galí, a fanciful alternate history, in Spanish (2009).[59]

The 2008 novel *Azazīl*, by Egyptian Muslim author Dr. Yūsuf Zaydan, tells the story of the religious conflict of that time through the eyes of a monk, including a substantial section on Hypatia;[60] Zaydan's book has been criticized by Christians in Egypt.[61]

Her life is portrayed in the Malayalam novel *Francis Itty Cora* (2009) by T. D Ramakrishnan.

Examples in English include

- *Remembering Hypatia: A Novel of Ancient Egypt* by Brian Trent,;[62]
- *Flow Down Like Silver, Hypatia of Alexandria* (2009) by Ki Longfellow, the second in a trilogy of the divine feminine, the first being *The Secret Magdalene*;[63]
- *The Plot to Save Socrates* (2006) by Paul Levinson and his sequel novelette "Unburning Alexandria" (2008) - where Hypatia turns out to have been a time-traveler from 21st Century America.

More factually, *Hypatia of Alexandria: Mathematician and Martyr* (2007) is a brief (113 page) biography by Michael Deakin, with a focus on her mathematical research. Hypatia has been considered a universal genius.[64]

The 2009 movie *Agora*, directed by Alejandro Amenábar, focuses on Hypatia's final years. Hypatia, portrayed by actress Rachel Weisz, is seen investigating the heliocentric model of the solar system proposed by Aristarchus of Samos, and even anticipating the elliptical orbits discovered by Johannes Kepler 1200 years later.

Notes

[1] Scholasticus, Socrates. *Ecclesiastical History* (http://cosmopolis.com/alexandria/hypatia-bio-socrates.html). .

[2] Krebs, *Groundbreaking Scientific Experiments, Inventions, and Discoveries*; *The Cambridge Dictionary of Philosophy, 2nd edition, Cambridge University Press, 1999:* "Greek Neoplatonist philosopher who lived and taught in Alexandria."

[3] Mueller, I.; L.S. Grinstein & P.J. Campbell (1987). *Women of Mathematics: A Biobibliographic Sourcebook*. New York: Greenwood Press.

[4] Columbia Encyclopedia, Hypatia (http://www.bartleby.com/65/hy/Hypatia.html) citation:*Alexandrian Neoplatonic philosopher and mathematician*

[5] Hypatia (http://www.britannica.com/eb/article-9041785/Hypatia), *Encyclopædia Britannica*: *"Egyptian Neoplatonist philosopher who was the first notable woman in mathematics."*

[6] Toohey, Sue (2003). "The Important Life & Tragic Death of Hypatia" (http://www.skyscript.co.uk/hypatia.html). Skyscript.co.uk. . Retrieved 2007-12-09.

[7] Hypatia of Alexandria – a philosophical martyr (http://www.abc.net.au/rn/philosopherszone/stories/2009/2530998.htm#transcript), *The Philosopher's Zone*. ABC Radio National (4 April 2009).

[8] Edward Jay Watts, (2006), *City and School in Late Antique Athens and Alexandria*. "Hypatia and pagan philosophical culture in the later fourth century" (http://books.google.com.br/books?id=0j94OFipg2MC&printsec=frontcover&hl=pt-BR&source=gbs_ge_summary_r&cad=0#v=onepage&q&f=false), pages 197–198. University of California Press

[9] "Women Philosophers in the Ancient Greek World: Donning the Mantle" (http://www.jstor.org/discover/10.2307/3810062?uid=3737664&uid=2&uid=4&sid=21101140085761), by Kathleen Wider. Hypatia © 1986 Indiana University Press p. 49–50;

Mangasarian, Mangasar Mugurditch. The Martyrdom of Hypatia, 1915
[10] Greenblatt, *The Swerve: how the world became modern* 2011:93.
[11] Christian Wildberg, in Hypatia of Alexandria – a philosophical martyr (http://www.abc.net.au/rn/philosopherszone/stories/2009/2530998.htm#transcript), *The Philosopher's Zone*, ABC Radio National (4 April 2009); Dzielska 1995, p. 105
[12] Michael Deakin (August 3, 1997). "Ockham's Razor: Hypatia of Alexandria" (http://www.abc.net.au/rn/science/ockham/or030897.htm). ABC Radio. . Retrieved 2008-05-23.
[13] Multicultural Resource Center: Hypatia (http://www.inventions.org/culture/female/hypatia.html)
[14] Dzielska 1995, p. 66
[15] *Historical Dictionary of Feminism*, by Janet K. Boles, Diane Long Hoeveler. p. 166.
[16] Suda online, Upsilon 166 (http://www.stoa.org/sol-bin/search.pl?db=REAL&search_method=QUERY&login=guest&enlogin=guest&user_list=LIST&page_num=1&searchstr=Hypatia&field=hw_eng&num_per_page=100)
[17] Bregman, J. (1982). "Synesius of Cyrene: Philosopher-bishop". Berkeley: University of California Press.
[18] Chronicle 84.87-103 (http://cosmopolis.com/alexandria/hypatia-bio-john.html)
[19] John, Bishop of Nikiu, Chronicle 84.87-103
[20] "Isidorus 1" entry in John Robert Martindale, (1980), *The Prosopography of the Later Roman Empire*. Cambridge University Press
[21] Kingsley, Charles. "Hypatia" preface agreeing with Gibbon quotation.
[22] Great Inspirations – Hypatia (http://www.abc.net.au/science/slab/inspire/hypatia.htm)
[23] A. Fitzgerald, Letters of Synesius of Cyrene, London, 1926. (Letter 154 of Synesius of Cyrene to Hypatia (http://www.webcitation.org/query?url=http://www.geocities.com/hckarlso/sletter154.html&date=2009-10-25+23:46:39)).
[24] Dzielska 1995, p. 28
[25] John, Bishop of Nikiu: *Chronicle*. London (1916). English Translation (http://www.tertullian.org/fathers/nikiu2_chronicle.htm)
[26] Socrates Scholasticus, *Ecclesiastical History*, born after 380 AD, died after 439 AD
[27] John of Nikiu, "Chronicle", c. 700 AD
[28] Ecclesiastical History,Bk VI: Chap. 15 (http://www.fordham.edu/halsall/source/hypatia.html)
[29] David Engels: *Zwischen Philosophie und Religion: Weibliche Intellektuelle in Spätantike und Islam*, in: D. Groß (Hg.), Gender schafft Wissen, Wissenschaft Gender. Geschlechtsspezifische Unterscheidungen Rollenzuschreibungen im Wandel der Zeit, Kassel 2009, 97-124. (http://www.upress.uni-kassel.de/online/frei/978-3-89958-449-3.volltext.frei.pdf)
[30] Dzielska 1995, pp. 71–2; "Until recently scholars thought that Hypatia revised Theon's commentary on *Almagest*. The view was based on the title of the commentary on the third book of *Almagest*, which read as follows: "Commentary by Theon of Alexandria on Book III of Ptolemy's *Almagest*, edition revised by my daughter Hypatia, the philosopher". Cameron, who has analyzed Theon's titles for other books of *Almagest* and for other scholarly texts of late antiquity, concludes that Hypatia corrected not her father's commentary but the text of *Almagest* itself. Thus, the extant text of *Almagest* could have been prepared, at least partly, by Hypatia".
[31] Grout, J. (n.d.). "Encyclopaedia Romana: Hypatia" (http://penelope.uchicago.edu/~grout/encyclopaedia_romana/greece/paganism/hypatia.html). . Retrieved 2009-05-16.
[32] Dzielska 1995, p. 72
[33] "Hypatia and Her Mathematics". *Mathematical Monthly* **101** (3): 234-243. March 1994.
[34] Ethlie Ann Vare and Greg Ptacek, Mothers of Invention (http://www.inventions.org/culture/female/hypatia.html) 1988, pp. 24–26.
[35] 'For the sake of completeness we must mention that fact that SYNESIOS in his letter to HYPATIA mentions a hydrometer, which according to some was already known in the fourth century AD to PRISCIANUS, that is a century before SYNESIOS and HYPATIA.', Forbes, 'A Short History of the Art of Distillation: from the beginnings up to the death of Cellier Blumenthal', p. 25 (1970).
[36] 'In 402, Hypatia receives a letter from the ailing Synesius giving a brief description of what he calls a hydroscope. This is a scientific instrument which was then in common use, although Hypatian is often credited with its invention.', Waithe, 'Ancient women philosophers, 600 B.C.-500 A.D.', p. 192 (1987).
[37] 'In his letters he describes a hydroscope (really a hydrometer) he has made as well as a catapult. In addition to that, he had constructed what the ancients called an astrolabe, an instrument that demonstrated celestial phenomena. Cicero describes one invented by Archimedes; Hipparchus had made another; two first century A.D. models were commemorated in the Greek Anthology;', Thomas, 'Paths from Ancient Greece', p. 69 (1988).
[38] 'It is generally accepted that Greek astrologers, in either the 1st or 2nd centuries BC, invented the astrolabe', Krebs, 'Groundbreaking Scientific Experiments, Inventions, and Discoveries of the Middle Ages and the Renaissance', p. 196 (2004).
[39] 'The invention of the astrolabe is usually attributed to Hipparchus of the second century BC. But there is no firm evidence to support this view. It is however certain that the instrument was well known to the Greeks before the beginning of the Christian era.', Sarma, 'The Archaic and the Exotic: studies in the history of Indian astronomical instruments', p. 241 (2008).
[40] Chris Marvin, Frank Sikernitsky The Window:Philosophy on the Web (http://www.trincoll.edu/depts/phil/philo/phils/hypatia.html)
[41] 'Claudius Ptolemy used astrolabes also, but Synesius says that his new one is an improved model based on later research.', Thomas, 'Paths from Ancient Greece', p. 69 (1988).
[42] "Ep. 15 is rather short, but gives interesting information: it contains a detailed description of a hydroscope which Synesius asks Hypatia to order for him in Alexandria, requesting that she herself oversee its construction.", Kari Vogt, "The Hierophant of Philosophy" - Hypatia of Alexandria, Kari Elisabeth Boerresen and Kari Vogt, Women's studies of the Christian and Islamic traditions: ancient, medieval, and Renaissance foremothers, p. 161 (1993).

[43] Synodicon, c. 216, in iv. tom. Concil. p. 484, as detailed in The History of the Decline and Fall of the Roman Empire, vol. 8, chapter XLVII
[44] Whitfield 1995, p.14
[45] Dzielska 1996, p. 18
[46] Dzielska 1995, p. 99
[47] Dzielska 1995 p. 67
[48] Ogilvie, M. B. (1986). Women in science: Antiquity through the 19th century. Cambridge, MA: The MIT Press.
[49] The History Of Hypatia, A most Impudent School-Mistress of Alexandria: Murder'd and torn to Pieces by the Populace, In Defence of Saint Cyril and the Alexandrian Clergy. From the Aspersions of Mr. Toland (http://www.polyamory.org/~howard/Hypatia/Lewis_1721.html).
[50] Dzielska 1995, p. 3
[51] Soldan, W.G. and Heppe, H., *Geschichte der Hexenprozesse,* Essen reprint 1990. p.82.
[52] http://fr.wikisource.org/wiki/Hypatie_%28Po%C3%A8mes_antiques%29
[53] http://fr.wikisource.org/wiki/Hypathie_et_Cyrille
[54] Edwards, Catharine. "Roman Presences: Receptions of Rome in European Culture, 1789–1945" pp. 112
[55] Snyder, J.M. (1989). *The woman and the lyre: Women writers in classical Greece and Rome.*. Carbondale, IL: Southern Illinois University Press.
[56] Marsh, Jan; Pamela Gerrish Nunn (1998). *Pre-Raphaelite Women Artists*. London: Thames & Hudson.
[57] "Hypatia of Alexandria: A woman before her time" (http://www.womanastronomer.com/hypatia.htm). The Woman Astronomer. November 11, 2007. . Retrieved 2007-12-03.
[58] "Hypatia Sans Pro, my new typeface" (http://blogs.adobe.com/typblography/2007/04/hypatia_sans.html). *Adobe Systems*. . Retrieved 2008-08-24.
[59] *Hypatia y la eternidad* Gali, Ramón ISBN 978-84-936773-9-8
[60] Rose al-Yusuf, May 9, 2008 reviewed by Arab-West Report, 2008, week 18, art. 48 (http://www.arabwestreport.info/)
[61] "The sin of apostasy in 'Azazīl'" (http://arabwestreport.info/awr/article_details.php?article_id=20102&ayear=2008&aweek=21&article_title=The sin of apostasy in â^cAzazÄ«lâ & article_t_date=May 24&article_p_date=2008-05-19&article_p_week=2008-05-25&t=s). Arab-West Report. 2008, Week 28, Art. 29. .
[62] "Remembering Hypatia: A Novel of Ancient Egypt" (http://www.rememberinghypatia.com/). .
[63] "Flow Down Like Silver, Hypatia of Alexandria" (http://flowdownlikesilver.com/). .
[64] MacDonald, Beverley and Weldon, Andrew. (2003). Written in Blood: A Brief History of Civilization (pg. 173). Allen & Unwin.

References

- Dzielska, Maria (1996) [1995]. *Hypatia of Alexandria*. trans. F. Lyra. Harvard University Press. ISBN 0-674-43776-4.
- edited by David Fideler. (1994) [1993]. Fideler, David. ed. *Alexandria 2: The Journal of Western Cosmological Traditions*. **II**. Phanes Press. ISBN 0-933999-97-6.
- Whitfield, Bryan J. (Summer, 1995). "The Beauty of Reasoning: A Reexamination of Hypatia and Alexandria" (http://math.coe.uga.edu/tme/issues/v06n1/4whitfield.pdf). *The Mathematics Educator* (University of Georgia) **6** (1): 14–21. Retrieved 2009-05-16.

This article incorporates text from a publication now in the public domain: Smith, William, ed. (1867). "article name needed". *Dictionary of Greek and Roman Biography and Mythology*.

Further reading

- Alic, Margaret (1986). *Hypatia's heritage : a history of women in science from antiquity through the nineteenth century*. Boston: Beacon Press. ISBN 0-8070-6731-8.
- Cameron, Alan; Long, Jacqueline (1993). *Barbarians and politics at the Court of Arcadius*. Berkeley: University of California Press. ISBN 0-585-13966-0.
- Deakin, Michael A. B. (Mar., 1994). "Hypatia and Her Mathematics" (http://www.maa.org/pubs/Calc_articles/ma055.pdf). *American Mathematical Monthly* (Mathematical Association of America) **101** (3): 234–243. doi:10.2307/2975600. Retrieved 2009-05-16.
- Kingsley, Charles (1853). *Hypatia, or New Foes with Old Faces* (http://www.gutenberg.org/ebooks/6308). Chicago: W.B. Conkley.
- Knorr, Richard (1989). *Textual Studies in Ancient and Medieval Geometry*. Birkhäuser. ISBN 0-8176-3387-1.

- Molinaro, Ursule (1990). "A Christian Martyr in Reverse: Hypatia". *A full moon of women*. New York: Dutton. ISBN 0-525-24848-X.
- Osen, Lynn M. (1990). *Women in mathematics*. Cambridge, MA: MIT Press. ISBN 0-262-15014-X.
- Teruel, Pedro Jesús (2011). *Filosofía y ciencia en Hipatia*. Madrid: Gredos. ISBN 978-84-249-1939-9.

External links

- Resources on Hypatia (http://www.polyamory.org/~howard/Hypatia): booklist, classroom activities
- Hypatia on ABC Radio (http://www.abc.net.au/rn/science/ockham/or030897.htm) Transcript of an interview with Dr Michael Deakin about his research on Hypatia, broadcast on Australia's ABC Radio National. Sunday, 3 August 1997
- International Society for Neoplatonic Studies (http://www.isns.us/)
- Hypatia, Alexandria's Great Female Scholar (http://www.smithsonianmag.com/specialsections/womens-history/Hypatia-Ancient-Alexandrias-Great-Female-Scholar.html): from Smithsonian magazine

Geminus

Geminus of Rhodes (Greek: Γεμῖνος ὁ Ῥόδιος), was a Greek astronomer and mathematician, who flourished in the 1st century BC. An astronomy work of his, the *Introduction to the Phenomena*, still survives; it was intended as an introductory astronomy book for students. He also wrote a work on mathematics, of which only fragments quoted by later authors survive.

Life

Nothing is known about the life of Geminus. It is not even certain that he was born in Rhodes, but references to mountains on Rhodes in his Astronomical works suggests that he worked there. His dates are not known with any certainty either. A passage in his works referring to the *Annus Vagus* (Wandering Year) of the Egyptian calendar of 120 years before his own time, has been used to imply a date of c. 70 BC for the time of writing,[1] which would be consistent with the idea that he may have been a pupil of Posidonius, but a date as late as 50 AD has also been suggested.[2]

The crater Geminus on the Moon is named after him.

Astronomy

The only work of Geminus to survive is his *Introduction to the Phenomena* (Greek: Εἰσαγωγὴ εἰς τὰ Φαινόμενα), often just called the *Isagoge*. This introductory astronomy book, based on the works of earlier astronomers such as Hipparchus, was intended to teach astronomy for beginning students in the subject. In it, Geminus describes the zodiac and the motion of the Sun; the constellations; the celestial sphere; days and nights; the risings and settings of the zodiacal signs; luni-solar periods and their application to calendars; phases of the Moon; eclipses; star phases; terrestrial zones and geographical places; and the foolishness of making weather predictions by the stars.[3]

He also wrote a commentary on Posidonius' work *On Meteorology*. Fragments of this commentary are preserved by Simplicius in his commentary on Aristotle's *Physics*.

Mathematics

Geminus also wrote extensively on mathematics, including a comprehensive *Doctrine, (or Theory) of Mathematics.*[4] Although this work has not survived, many extracts are preserved by Proclus, Eutocius, and others. He divided mathematics into two parts *Mental* (Greek: νοητά) and *Observable* (Greek: αἰσθητά), (or in other words, Pure and Applied.) In the first category he placed geometry and arithmetic (including number theory), and in the second category he placed mechanics, astronomy, optics, geodesy, canonics (musical harmony), and logistics. Long extracts of his work are also preserved by Al-Nayrizi in his commentary on Euclid's Elements.

Notes

[1] Dicks, D., *Dictionary of Scientific Biography.* New York. (1970).
[2] Neugebauer, O., *A History of Ancient Mathematical Astronomy.* New York. (1975).
[3] Evans, J., *The History and Practice of Ancient Astronomy*, page 91. Oxford University Press. (1998).
[4] Heath, T., A Manual of Greek Mathematics, Dover Publications. (2003).

Bibliography

- Evans, J.,, Berggren, J.L., *Geminos's Introduction to the Phenomena: A Translation and Study of a Hellenistic Survey of Astronomy.* (http://press.princeton.edu/titles/8330.html) (Princeton University Press, 2006.) (This is the first complete English translation of this book.)

External links

- Εἰσαγωγὴ εἰς τὰ Φαινόμενα (Introduction to Phaenomena), original text online (http://www.astrologicon.org/geminus/geminus-intro.html)
- Technology Museum of Thessaloniki Entry (http://www.tmth.edu.gr/en/aet/2/49.html)
- O'Connor, John J.; Robertson, Edmund F., "Geminus" (http://www-history.mcs.st-andrews.ac.uk/Biographies/Geminus.html), *MacTutor History of Mathematics archive*, University of St Andrews.
- PDF scans of Manitius' edition of the Geminus' *Elementa Astronomiae* ("Elements of Astronomy") - public domain (http://www.wilbourhall.org) (Classical Greek with German translation)]

Hippocrates_of_Chios

Hippocrates of Chios was an ancient Greekmathematician, (geometer), and astronomer, who lived c. 470 – c. 410 BCE.

He was born on the isle of Chios, where he originally was a merchant. After some misadventures (he was robbed by either pirates or fraudulent customs officials) he went to Athens, possibly for litigation. There he grew into a leading mathematician.

On Chios, Hippocrates may have been a pupil of the mathematician and astronomer Oenopides of Chios. In his mathematical work there probably was some Pythagorean influence too, perhaps via contacts between Chios and the neighbouring island of Samos, a center of Pythagorean thinking: Hippocrates has been described as a 'para-Pythagorean', a philosophical 'fellow traveler'. The *Reductio ad absurdum* argument (or proof by contradiction) has been traced to him.

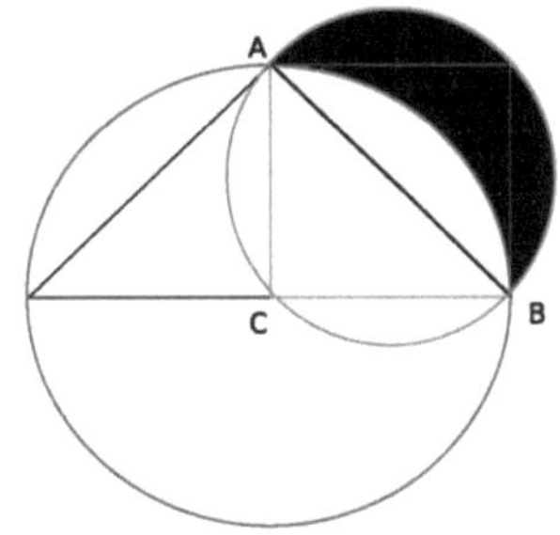

The Lune of Hippocrates. Partial solution of the "Squaring the circle" task, suggested by Hippocrates. The area of the shaded figure is equal to the area of the triangle ABC. This is not a complete solution of the task (the complete solution is proven to be impossible with compass and straightedge).

Mathematics

The major accomplishment of Hippocrates is that he was the first to write a systematically organized geometry textbook, called *Stoicheia Elements*, that is, basic theorems, or building blocks of mathematical theory). From then on, mathematicians from all over the ancient world could, at least in principle, build on a common framework of basic concepts, methods, and theorems, which stimulated the scientific progress of mathematics.

Only a single, and famous, fragment of Hippocrates' *Elements* is existent, embedded in the work of Simplicius. In this fragment the area is calculated of some so-called *Hippocratic lunes* — see Lune of Hippocrates. This was part of a research programme to achieve the "quadrature of the circle", that is, to calculate the area of the circle, or, equivalently, to construct a square with the same area as a circle. The strategy apparently was to divide a circle into a number of crescent-shaped parts. If it were possible to calculate the area of each of those parts, then the area of the circle as a whole would be known too. Only much later was it proven (by Ferdinand von Lindemann, in 1882) that this approach had no chance of success, because the factor pi (π) is transcendental. The number π is the ratio of the circumference to the diameter of a circle, and also the ratio of the area to the square of the radius.

In the century after Hippocrates at least four other mathematicians wrote their own *Elements*, steadily improving terminology and logical structure. In this way Hippocrates' pioneering work laid the foundation for Euclid's *Elements* (c. 325 BC) that was to remain the standard geometry textbook for many centuries. Hippocrates is believed to have originated the use of letters to refer to the geometric points and figures in a proposition, e.g., "triangle ABC" for a triangle with vertices at points A, B, and C.

Two other contributions by Hippocrates in the field of mathematics are noteworthy. He found a way to tackle the problem of 'duplication of the cube', that is, the problem of how to construct a cube root. Like the quadrature of the circle this was another of the so-called three great mathematical problems of Antiquity. Hippocrates also invented the technique of 'reduction', that is, to transform specific mathematical problems into a more general problem that is more easy to solve. The solution to the more general problem then automatically gives a solution to the original problem.

Astronomy

In the field of astronomy Hippocrates tried to explain the phenomena of comets and the Milky Way. His ideas have not been handed down very clearly, but probably he thought both were optical illusions, the result of refraction of solar light by moisture that was exhaled by, respectively, a putative planet near the sun, and the stars. The fact that Hippocrates thought that light rays originated in our eyes instead of in the object that is seen, adds to the unfamiliar character of his ideas, which belong to the realm of speculative philosophy.

Others

Under the name of Isicrate, Alfred Jarry attributed to him the origins of 'pataphysics.

References

- Ivor Bulmer-Thomas, 'Hippocrates of Chios', in: *Dictionary of Scientific Biography*, Charles Coulston Gillispie, ed. (18 Volumes, New York 1970–1990) pp. 410–418.
- [Axel Anthon] Björnbo, 'Hippokrates', in: Paulys Realencyclopädie der Classischen Altertumswissenschaft, G. Wissowa, ed. (51 Volumes; 1894–1980) Vol. 8 (1913) col. 1780–1801.

External links

- O'Connor, John J.; Robertson, Edmund F., "Hippocrates of Chios" [1], *MacTutor History of Mathematics archive*, University of St Andrews.
- The Quadrature of the Circle and Hippocrates' Lunes [2] at Convergence [3]

References

[1] http://www-history.mcs.st-andrews.ac.uk/Biographies/Hippocrates.html

[2] http://mathdl.maa.org/convergence/1/?pa=content&sa=viewDocument&nodeId=1203&bodyId=1593

[3] http://mathdl.maa.org/convergence/1/

Article Sources and Contributors

Serenus_of_Antinouplis *Source*: http://en.wikipedia.org/w/index.php?title=Serenus_of_Antinouplis *Contributors*: Ceroklis, Delirium, Eklektro68, NorwegianBlue, Oracleofottawa, PasabaPorAqui, Rich Farmbrough, Sam Blacketer, Selfworm, WestwoodMatt, 6 anonymous edits

Hellenistic_civilization *Source*: http://en.wikipedia.org/w/index.php?title=Hellenistic_civilization *Contributors*: 2help, Abutorsam007, Adam Bishop, Aksarajean, Al1encas1no, Alagos, Alansohn, Aldux, Alsandro, AndrewHowse, Angelamurock, Ashmedai 119, Berean Hunter, Bjankuloski06en, Bloublou, BoH, Boccobrock, Bongwarrior, Bontekoe, Brando130, Calienfj, Careshi, Catalographer, Chaleyer61, Cole54355, CommonsDelinker, Coppertwig, Corusant, Cribananda, CuteHappyBrute, Davidiad, Dbachmann, Dedede101, Denis Bilodeau, Deor, Dickdock, DocWatson42, Draganparis, EdJohnston, Editor2020, Editorinfo, Ehn, Elmondo21st, Erud, Ewawer, Felicitycobb, FrancoGG, Gakrivas, Garion96, Geologyguy, Glacialfox, Hectorian, HiMyNameIsFrancesca, Hmains, I dream of horses, Inferno, Lord of Penguins, J.delanoy, JDP90, Jab843, Jagged 85, Javierfv1212, Jeff3000, Jenks24, JoanneB, Joel7687, Joriki, Jschnur, Jyusin, Ka Faraq Gatri, Khazar, Kirobos, Krakkos, Logout here, MONGO, Macedonian, Macintosh User, Marek69, Markussep, Mattis, Mild Bill Hiccup, Minimac's Clone, MinisterForBadTimes, Modest Genius, Natalie Erin, Nesnad, Nev1, Orphan Wiki, Osagab, Parent5446, Parthian Scribe, Paul August, Per Honor et Gloria, Peripatetic, Poetaris, Reddi, Rich Farmbrough, SamEV, Seanwall11111, SimonP, SirBob42, Sowlos, Spicemix, Stephan Leeds, Stevepeterson, Suwa, Tbarker, Techman224, Treezone400, Wahrmund, Welsh, Wikiklrsc, Wimt, Woohookitty, Yachtsman1, Yom, 194 anonymous edits

Mathematician *Source*: http://en.wikipedia.org/w/index.php?title=Mathematician *Contributors*: 1exec1, 203.26.98.xxx, 28421u2232nfenfcenc, 63.192.137.xxx, A little insignificant, A4, Aarsalankhalid, Abhishekitmbm, Addshore, Ahoerstemeier, Ajraddatz, Akjar13, Alansohn, AlexGTV, Alexandria, Almit39, Alterationx10, Ancheta Wis, Andmatt, Andre Engels, Animum, Anir1uph, Anonymi, Anonymous Dissident, Anonymous the Editor, Antandrus, Ante Aikio, Anthony Appleyard, Anton Gutsunaev, Apptas, Aqwis, Architectual, Arjun01, Avoided, AxelBoldt, BD2412, Babalapaloop, Banej, BarretB, Bassbonerocks, Bbggae, Belovedfreak, Bevo, Bfigura's puppy, Bhawani Gautam, Bhny, BlaiseFEgan, Bliduta, Blinking Spirit, Blotwell, Bobo192, Bongwarrior, Bootyy, Booyabazooka, Brandmeister (old), Brianjd, Brion VIBBER, Bte99, ButOnMethItIs, Buzybeez, C S, C.Fred, CBDunkerson, CBM, CRGreathouse, Cafeturque, Calabe1992, Calypso, CambridgeBayWeather, Camilo Sanchez, Canderson7, Canuckistan Rocks, Capitalist, Catmoongirl, Cenarium, Cgtdk, Chapeaubien, Chas zzz brown, Chewings72, ChicXulub, ChrisGualtieri, Chrisbehk, Ciphers, Cj67, Cliff12345, Closedmouth, CommonsDelinker, Conditionhumaine, Conversion script, Coolman2k8, Correogsk, Cpl Syx, CptCutLess, Cronholm144, D.M. from Ukraine, DARTH SIDIOUS 2, DBigXray, DSachan, Damirgraffiti, Daniel Case, Darnell24, Darolew, David Eppstein, David.Monniaux, DavidLevinson, Db099221, Dbenbenn, Dbroadwell, Dcclark, Dcljr, DealPete, Demmy100, DerHexer, Derickjay, Deus Ex, Deville, Dicklyon, DigitalCharacter, Dlohcierekim, Docu, Dominus, DonaNobisPacem, Donarreiskoffer, Doulos Christos, DoversCHAMP, Drmies, Drosdaf, Dsgdfshfdshdsfh, Eastlaw, Eclecticology, Edcolins, Edison, Eeekster, Elockid, Epbr123, Epsilon60198, Eric119, Etale, Eternityflow, Everyking, Excirial, FF2010, FJPB, Falcon8765, Favonian, Fephisto, Foljiny, Fredrik, Frentz, Fyyer, Fæ, GTBacchus, Galoubet, Gap9551, Garald, Garycompugeek, Gauss, Geneffects, Georg Muntingh, Gguidelas, Ghoshsanjoy, Giftlite, Gilliam, Gioto, Glane23, Gogo Dodo, Goodnightmush, Googl, Gpvos, Grafen, Graham87, Grammar conquistador, Greatuser, Greg Kuperberg, Grim23, Grimmcar, Grinder0-0, Gurchzilla, Haham hanuka, HamburgerRadio, Hdt83, HedgeHog, Helder.wiki, HelloAnnyong, Hemmingsen, Hennessey, Patrick, Henrik, Heracles31, Hu12, I dream of horses, ICanAlwaysChangeThisLater, Iain99, Ianw17, Ibbn, Igoldste, IndieAddiction, Insanity Incarnate, Iothiania, Iriseyes, Irrªtiºnal, Ixfd64, J'88, J.delanoy, J271828, JDspeeder1, JForget, Jackfork, Jackol, Jacobolus, Jagged 85, Jahiegel, Jaimeastorga2000, Jallotta, JamesMLane, Jamesofur, JamieJones, Jankur, Janlo, Jcm, Jeff G., Jeffrey Mall, Jeffrey O. Gustafson, Jeffroyracer, Jeronimo, Jfhgja;, JillandJack, Jim.belk, JimWae, Jitse Niesen, Joe Fabeetz, Joeldudesx2, Joerite, John254, Jonas Mur, Joseph Solis in Australia, Joshua Issac, JoshuaZ, Jstnmgr, Juan Marquez, Juliancolton, Jusdafax, Justin W Smith, Jvalure, Kaldari, Karada, KarlHallowell, Katherine, Kazikame, Kehrbykid, Kiefer.Wolfowitz, Kinaro, Kindyin, King of Hearts, Kingbusch, Kkm010, Koavf, Konstable, Koyaanis Qatsi, Kubigula, Kudret abi, Kungfuadam, Kungming2, Kuru, LA2, LJosil, Laker9903, Lambiam, Leon math, Leuko, Lihaas, Little guru, LittleDan, Logicalasif, Lradrama, Lugia2453, Lupin, Lupo, Lyleq, LéonTheCleaner, M.O.X, MKoltnow, Madhero88, Maelnuneb, Maidenindigo, Mannafredo, Manning Bartlett, Marcika, Marek69, Master of Puppets, Masterpiece2000, Materialscientist, MathKnight, Mathmo, Matthead, Maurice Carbonaro, MegaPedant, Menareretards, Meni Rosenfeld, Meno25, Mets501, Mhym, Michael Hardy, Michael Slone, Micione, Mikez, Minimac's Clone, Miquonranger03, Mir76, Misza13, Mmnaw, Moderate2008, Moe Epsilon, Morios, Mossgiantkiller, Moverton, Mrdthree, Mtking, Myasuda, Myopic Bookworm, Mysteronald, N5iln, Nburden, Nbvjou, Nepenthes, Netoholic, Nev1, Newkai, Nick 1234567, NikolaiLobachevsky, Nixdorf, Nuadh, Nwwaew, Ocaasi, Ocolon, Oleg Alexandrov, Olivier, Omicronpersei8, Orange Suede Sofa, Out of Phase User, Oğuz Ergin, PMajer, Palakisbest, Pan Dan, Patrick, Patrick Berry, Paul August, Paw7151, Paweł Ziemian, Pdcook, Persian Poet Gal, Pete.Hurd, Peterlin, Petrb, Pgecaj, PhS, Phantomsteve, Philip Trueman, Piano non troppo, Pichu826, Pinethicket, Pleasantville, Point-set topologist, Poor Yorick, Poorleno, Pred, Pricebank, Prolog, Proofreader77, Prunesqualer, Psychonaut, Pupster21, Quentar, Quuxplusone, R. fiend, RDBury, RMcGuigan, Raepetersen, Reach Out to the Truth, Reddi, Redfarmer, Redman775, Reedy, Requestion, Rhythm, Rich Farmbrough, Richard L. Peterson, RjLesch, RobLa, Robert W. Wright, RobertG, Robinh, Ronhjones, Rotem Dan, Rrokkedd, Rsrikanth05, Ryulong, SCZenz, SU Linguist, Sam Korn, Samuel, Samuel Bach, Sandman, Sarenne, Satori Son, Schneelocke, SchoolPlayAround3, SchuminWeb, Scohoust, Sdornan, Seaphoto, Semperf, Sepastaj, Shadowjams, Shanes, Shimmin, Shoeofdeath, Shogartu, Sibian, SimonP, Sinadoom, Sj, Smalljim, Sp33dyphil, SpuriousQ, Srich32977, Srisou, StaticGull, Stephenb, Strojo4, Suklaa, Sushiflinger, SusikMkr, Synchronism, Syrusakbary, TakuyaMurata, Tangent747, Tawker, Taxman, Tbackstr, Teeth123123, Terrek, Tesseran, Tetraedycal, The Anome, The Fish, The Thing That Should Not Be, Themanofthehouserulzzz, Therepel, Thingg, Tide rolls, TigerShark, Tim1357, Timwi, Titodutta, To Fight a Vandal, Tobias Bergemann, Toddst1, Tom Morris, Tomgally, Tomgreeny, Tommy2010, Tony pal, TonyBallioni, TooMuchMath, Topology Expert, Tosayit, Trang Oul, Typuifre, Ukexpat, Ulric1313, UltimateKingSatan, Uncle Dick, Uncool95, UnitedStatesian, Useight, Vanished User 1004, Vanished user 342562, Versus22, Vertium, Viriditas, Vishnava, Vornoff, Vrenator, Vsmith, Vytautas, Wafulz, Waltervulej, Wayward, Weerasad, Whaa?, Why Not A Duck, WikHead, WikipedianMarlith, Wikipelli, Wilanthule, William Avery, Winchelsea, Wine Guy, Wlod, Wolfrock, Wtmitchell, Wuffyz, Wwannsda, Wysprgr2005, XJamRastafire, Xp54321, Yamamoto Ichiro, Yankeetown12, Ybbor, Yfkmafiaboss, Z3dman, ZX81, Zmoney918, Zomno, ZooFari, Zsinj, Zundark, Zvika, Zzyxzaa26, Александър, 1170 anonymous edits

Theon_of_Smyrna *Source*: http://en.wikipedia.org/w/index.php?title=Theon_of_Smyrna *Contributors*: Catalographer, Chargerz09, EWS23, FeanorStar7, Fratrep, Ghirlandajo, Giftlite, Grsalmonerous, KRBN, Loupeter, Mindspillage, Oldtimereligion, Omassey, Roger Pearse, Samsam9, Sceptic1954, Selfworm, Singinglemon, Syncategoremata, TheOldJacobite, WestwoodMatt, XJamRastafire, Ziggurat, 3 anonymous edits

Hadrian *Source*: http://en.wikipedia.org/w/index.php?title=Hadrian *Contributors*: A Little Bit of Liberace, A little insignificant, A3RO, Achillestendon69, Adam Bishop, Adam Keller, Adam114, Adambro, Addshore, Adranus, Ajunited, Alan012, AlexiusHoratius, Alicia M. Canto, Alteripse, And we drown, Andre Engels, Andrei nacu, Andrew Gray, Andyluciano, Angela, Anglius, Angusmclellan, Anna Lincoln, Anriz, Antandrus, Anthony Bradbury, Aramgar, Arda Xi, ArglebargleIV, Arkuat, Armeria, Arnoutf, Arthurian Legend, Asarelah, Astynax, Attilios, Aurelius Victor, Auró, Avram Fawcett, Awe57a347adhw34a67a3578am356m8648, BD2412, Barbatus, Bastin, Bearcat, Beeblebrox, Ben Babcock, Ben-Zin, Biker Biker, Bill Thayer, Bje2089, Blahm, BloodDoll, BoBo, Boarder dude21, Bobo192, Boing! said Zebedee, BongHitz4Musa, Bozhawk, BraneJ, Brian Crawford, Brianyoumans, Bubba hotep, Budija, Bulgaroctonus, Burntsauce, Byox, CJGB, CWii, CaesarInvictus, Caltas, CambridgeBayWeather, CanadianLinuxUser, Captain panda, Captain-n00dle, Catalographer, Catiline63, Cdc, Cenedi, Censusdata, CharlotteWebb, Che829, Chicago-Ruless, ChrisO, ChristopherWillis, Clancyolincy, Clio the Muse, Closedmouth, Cnyborg, Codex Sinaiticus, Coffee, CommonsDelinker, Condem, Conversion script, Corax, Cottingham, Crimsonmargarine, Crotalus horridus, Cst17, CynicalMe, Cynwolfe, D-Katana, D. Webb, DabMachine, Daisykc, Daizus, Dakirw8, DanB DanD, Danilot, Danny, DannyDaWriter, Dave souza, Davehi1, Davidiad, Deb, Dejvid, Delconvento, Den fjättrade ankan, Denkealsobin, Dersonlwd, Deucalionite, Dimadick, Dionysiaca, Discospinster, Dobie80, Doc glasgow, Don4of4, Donjanssen, DoubleBlue, Dougweller, Dppowell, Dpwkbw, Dr.K., DrKiernan, Drolsi Susej, Dvl007, Dysepsion, EALacey, EWS23, EauLibrarian, Echerbony, Eclecticology, EdJohnston, Edward, El C, Enviroboy, Epbr123, EraNavigator, Estudiarme, Everyking, Extramural, FF2010, Fano Ksenaki, Fatidiot1234, Fatman999999, Fdewaele, FeanorStar7, Fedallah, Felix Folio Secundus, Ferkelparade, Fieldday-sunday, FilipeS, Filpaul, FisherQueen, Fleabox, Fmandcac, Fordmadoxfraud, Fram, Francesca Santoro, Frankenpuppy, Freebirdthemonk, Fritz28408, Fruityjon, Fryslan0109, Fulcher, Fæ, G.-M. Cupertino, G.W., Gaius Cornelius, Galassi, Garcilaso, Gareth E Kegg, Gdr, GeckoRoamin, GeneralPatton, Genesisflare, Geriech, Gfoley4, Ghaly, Ghepeu, GideonF, Gilliam, Ginsengbomb, Giovanni33, Girlfawkes, Glanns, Gogo Dodo, Grafen, GrandpaHan, GreatWhiteNortherner, Greco22, Grim23, Ground Zero, Gsrgsr, Guardianofhistory, Gugganij, Gurchzilla, HDCase, Hadal, Hadrian89, Haiduc, HalfShadow, Hall Monitor, HamburgerRadio, Haploidavey, HarryHenryGebel, Hede2000, Henrik, Hmains, Hmzacarias, Hu12, Huangdi, Humus sapiens, Hut 8.5, IZAK, Ian.thomson, Icarus3, Ida Shaw, Iglonghurst, Imbor3d, Invictus84, Iridescent, Isis, J.delanoy, J04n, JASpencer, JHunterJ, JRPG, JaGa, Jack12809, Jack1956, James086, JamesBWatson, Jamyskis, Jan1nad, Jaraalbe, Jclemens, Jeronimo, Jfurr1981, Jlittlet, Joao 1000, Joe Hepperle, John K, John254, Jonecc, Josep Azuara, Jschubart, Just H, Jyril, Kaisershatner, Karlthegreat, Kars777, Kauffner, Kbdank71, Kbh3rd, Kbolino, Keegan, Kevin.cohen, Khazar2, Kingpin13, Kingturtle, Kiomarv, KitMarlowe2, KnowledgeOfSelf, Koyo123, Krumpinjugger, Kubigula, Kuralyov, Kurt Leyman, LOL, La convivencia, Lars T., Lerdthenerd, LibraryLion, Lidaye, Lightmouse, Lionelt, LizardJr8, Llywrch, Logan, Looxix, Lord Emsworth, Lord Xuan, Lothar von Richthofen, Lotje, Lovemyrice, Luna Santin, Lupus Bellator, MBisanz, MDCore, MER-C, MONGO, MPF, Macrocaus, Madmagican3, Magister Mathematicae, MakeRocketGoNow, ManyFireflies, MarcK, MartinHarper, Matthew191, Mattis, Mav, Maxis ftw, Mbc362, Metb82, Meursault2004, Michael Hardy, MichaelTinkler, Mimihitam, Minesweeper, Mirv, Mlouns, Mmccalpin, Monegasque, Morn, Mr. Random, MrFish, Muriel Gottrop, Myanw, Mygerardromance, MysteryDog, Nach0king, NantonosAedui, Nasz, Ndick5913, Neddyseagoon, Neoaeolian, Nergaal, Nev1, Nicke L, Nkocharh, Nrets, Odin1, Odysses, Ojigiri, Olivier, Olorin28, Onco p53, Oosh, Optimist on the run, Oxymoron83, P Aculeius, Panairjdde, Part Deux, Pascal yuiop, Paul August, Paul Barlow, Pepefe, Pernoctator, Pewwer42, Pgk, Philip Trueman, PiCo, Piccadilly, Pinball22, Pince Nez, Pjthompso, Poliocretes, Polylerus, Ponyo, Possum, Pring, Puffin, Pvasiliadis, Qohen, QuartierLatin1968, R'n'B, RFB, Radioflux, Rasiel, Raymond Cruise, Razorflame, Rbyteisbst, Rcpaterson, Redmarkviolinist, Regibox, Retired username, Reywas92, Rhodian, Rich Farmbrough, Richard Arthur Norton (1958-), Richard75, Rikimaru, Rlest, Rms125a@hotmail.com, RomanSpa, Ronhjones, Ronincyberpunk, Rorschach, Rover1337, Royalty Parsing Project, Rumiton, Rursus, Ryanthompson, Ryuko2001, S1rkull, S711, SSSN, Salmar, SanGavinoEN, Saperaud, Sapiens23, Sardonac, Senjuto, Sfodero, Shanes, Shawisland, Shoaler, Siafu, Silverhorse, Sionus, Sleigh, SlightlyMad, Slon02, Slowking Man, Smack, SmilesALot, Snowolf, Some jerk on the Internet, Special-T, Spencer, Spino, Stan Shebs, StaticGull, Steerpike, Steven Zhang, Stevenmitchell, Stevensaylor, Str1977, Stwalkerster, Sun Creator, Surtsicna, THEN WHO WAS PHONE?, Tannin, Tarquin Binary, Tataryn77, Tcncv, Teemu Ruskeepää, Tetraktys-English, Th4n3r, The Interloafer, The Ogre, The Rambling Man, The Stickler, TheRanger, TheRegicider, Thecheesykid, Theforcecanspeak, ThievingBeagles, Thingg, Thismightbezach, Thryduulf, Tim!, Timrollpickering, Tommy2010, Tomo64, Tpbradbury, Travelbird, Truco, Tubby5555, Twospoonfuls, Tynetrekker, UkPaolo, Useight, Vahagn Petrosyan, VampWillow, Vanished user ewfisn2348tui2f8n2fio2utjfeoi210r39jf, Velvetron, Vespuccia, Vilartatim, Vsb, Vulcanstar6, WAvegetarian, Waggers, Wetman, White Shadows, WiKiller, Widr, Wikignome0529, Wikihistorian, Wikipelli, Woohookitty, World, XJTrecoX, Y(J)S, Yintan, Yonmei, Zarateman, Zenzee, Zoicon5, Zsero, Zundark, 1011 anonymous edits

Conic_section *Source*: http://en.wikipedia.org/w/index.php?title=Conic_section *Contributors*: 21655, 24.191.126.xxx, Academic Challenger, Adilch, Aiden Fisher, Aitias, Al-Andalusi, Aldaron, Almit39, Andre Engels, Andrw1491, Anil1956, Anonymous Dissident, Armend, Army1987, Arthena, Artie p, Asdfdsa, Assassinoc714, AugPi, AxelBoldt, Barsamin, BaseballDetective, BasilRazi, Bbogren, BenFrantzDale, Bender235, Blondandy, Bomazi, Brian0918, CBM, CRGreathouse, Calvin 1998, Canadianb123, Caramdir, Catgut, Ceroklis, Charles Matthews, Cheeser1, Chenyuntao, Christian Thiemann, Christopher Thomas, Chuckrocks, Cmapm, Colin Rowat, Colonies Chris, Conversion script, Crazytales, Croikle, DMacks, DVdm, Dangerousnerd, Danheeks, Daniel Verity O'Connor, Danny, Daqu, Davercox, David Pierce, DavidHouse, Dbenbenn, Dcoetzee, DeadEyeArrow, DerHexer, Dhexus, Dino, Discboy, Discospinster, Doctormatt, Dominus, Dthomsen8, Duk, Duoduoduo, Durin, Dvd Avins, EL Willy, ELApro, ERuede, Ellywa, Elmer Clark, Enchanter, Eric119, Error792, EscapingLife, Feyrauth, Filpaul, Flashmorbid, Fropuff, FrozenMan, Galadh, Gene Ward Smith, Geometryfan, Giftlite, GilbertoSilvaFan, Gilliam, GoingBatty, Grabba, Greensburger, Gtstricky, Hager e, Halo, Hannes Eder, Harcalion, Herbee, Hfastedge, Hgilbert, Hut 8.5, Ideyal, ImMAW, ImperatorExercitus, InkKznr, Inquam, InverseHypercube, Ivanmuratov, J.delanoy, JCSantos, JForget, Jackzhp, JamesAM, JamesBWatson, January, Jarekt, JelmerW, John of Reading, JohnWittle, Jonpro, Jorge Stolfi, Jtico, Juansempere, Kajisol, Knucmo2, Knutux, LOL, Laur2ro, Life, Liberty, Property, Lir, Looxix, Lumos3, Manop, Martinp23, Meol, Merphant, Michael Hardy, Michaelliv, Mike Capp, Mikm, Mmeijeri, Modelpanicer, Morana, MrOllie, Mrjohncummings, Mwtoews, Natsirtguy, NawlinWiki, Nbarth, Ncurses, NewEnglandYankee, Newone, Nic bor, Nick.sideras, Nickptar, Nk, Nonagonal Spider, Nuno Tavares, Oleg Alexandrov, OwenX, Ozgod, Paisa, Patrick, Paul August, Paulnwatts, Pbroks13, Peruvianllama, Phancy Physicist, Philmac, Pigei, Pokipsy76, Pushhigher, Pwaldron, Qfl247, Quadell, R'n'B, RDBury, Radagast83, Rausch, Raven in Orbit, Rb88guy, Rdsmith4, Rgdboer, Riojajar, Rmt2m, RobHar, Rumping, Rvollmert, SEIBasaurus, Salix alba, Sam Derbyshire, SammyD123, Schizobullet, Schmalx, Siddhant, SiegeLord, Silly rabbit, SpaceFlight89, SpongeSebastian, Srinivas.zinka, StradivariusTV, StuRat, Sverdrup, Swift, Tarquin, Thatotherdude, Thingg, Tiggerjay, Tkuvho, Tobias Bergemann, Tosha, Tran Quoc123, Tsingi, Typometer, Urhixidur, VegaDark, Vincnet, Wapcaplet, Warut, Wcherowi, Wessonality, Whommighter, Whoop whoop pull up, Wknight94, Wmayner, Woofles, Woohookitty, Wwoods, XJamRastafire, Yamamoto Ichiro, Yekrats, Yorihs, Yosha, Zundark, Վազգեն, 280 anonymous edits

Apollonius_of_Perga *Source*: http://en.wikipedia.org/w/index.php?title=Apollonius_of_Perga *Contributors*: 1297, Adamsan, Almit39, AndyHuston, Anlace, Art LaPella, BillLoney, Bináris, Brad7777, C.Fred, Capricorn42, CarolGray, Casey Abell, Catalographer, Cbustapeck, Cenarium, Charles Matthews, ChicXulub, DO'Neil, DanMS, David Eppstein, Dogaru Florin, Dr. Submillimeter, Duoduoduo, EALacey, EauLibrarian, Ego White Tray, Englejef, FeanorStar7, Fraggle81, Func, Geboy, Giftlite, Glacialfox, Granite07, Herbee, Hgilbert, Icairns, Ivan Štambuk, JYOuyang, Jack Greenmaven, Jan.Kamenicek, Japanese Searobin, Jaredwf, Jessemerriman, Jgeortsis, Jgtl2, Josephk, Joy, Jumbuck, Jusdafax, Ka Faraq Gatri, Kupaka, Lapisphil, Lzur, Maestlin, Magnus Manske, Maksym Ye., Maximus Rex, Mfried60, Mhaitham.shammaa, Michael Hardy, Mike Christie, Mr.Z-man, MrOllie, Munita Prasad, Mushroom, Nic bor, Nicke L, Obradovic Goran, Omnipedian, Ouedbirdwatcher, PBS-AWB, Pearle, Phoebe, Pmanderson, Polylerus, Pwqn, Quotemstr, RDBury, Ragesoss, Raverdrew, Retired username, Rjwilmsi, RogDel, Rojypala, S.K., Salix alba, Sam nead, Scrawlspacer, Seaworld, Sedonaarizona, Selfworm, Singinglemon, Skeptic2, Sonett72, Spacepotato, Spalding, Spiritia, Stbalbach, Stevertigo, Struway, Syncategoremata, Telfordbuck, Tkuvho, Tom Peters, Ttwo, Varlaam, Veddharta, Vividupperer, Wareh, WikiPeterD, Wikijens, WillowW, XJamRastafire, Yakudza, Yohan (China), Zaheen, Ævar Arnfjörð Bjarmason, , 107 anonymous edits

Archimedes *Source*: http://en.wikipedia.org/w/index.php?title=Archimedes *Contributors*: 21655, A Softer Answer, A spiros, A. di M., APT, Aaron Schulz, AaronS, AbhijayM., Abhishikt, Acalamari, Acroterion, Adamarthurryan, Adashiel, Admrboltz, Adrionwells, AdultSwim, Aetheling, Afed, AgnosticPreachersKid, Agutie, Ahoerstemeier, Aitias, Albanaco, Aldux, Alex Hytha, Alex.muller, Alexbrewer, AlexiusHoratius, Alison, AllanBz, Amalthea, An Siarach, Ancheta Wis, Andre Engels, Andres, Andrew Gray, Andrew Norman, Andy Ross, Anetode, Angelic Wraith, Anicholo, Animum, Anitalittleone2, Anonymi, Anonymous Dissident, Anskas, Antandrus, Antnyp15, Aratuk, Arcfrk, Archimedes lover, Archimerged, Archmagi1, Argos'Dad, ArielGold, Arthena, Arthursimms, Astrognash, Astronomyinertia, Asyndeton, Atelaes, Atsushi2, Attilios, AuburnPilot, Austy06, Average Earthman, Avicennasis, AxelBoldt, B0b Jones, Badgernet, Bananainmypants4u, Baudrillard, Bearingbreaker92, Ben-Zin, Bender235, Bhadani, Big Bird, Bigmak, Bigturtle, BillC, Binabik80, Blankfaze, Blockmyass, Bluemask, Bobo192, Boldra, Boleyn, Bongwarrior, Bonzai273, BorgHunter, BozMo, Brandmeister, Brewers49, Brianlucas, BrightStarSky, Brighterorange, Brtkrbzhnv, Bruce1ee, Brutaldeluxe, Bubba hotep, Bubba73, Bytwerk, CBM, CRGreathouse, CWY2190, CWii, Caffelice, Caltas, Can't sleep, clown will eat me, Canadian-Bacon, CanadianCaesar, Canonblack, Canuckle, Captain panda, Carabinieri, Catalographer, Catgut, Centrx, Cfailde, Charles Matthews, Chazsivad, Cherryalexandra, Cholmes75, Choster, Chris Chittleborough, Chrislk02, Chrisobyrne, ChristopherWillis, Chun-hian, Cimon Avaro, ClanCC, Closson the cleaner, Cometstyles, CommonsDelinker, Connormah, Conone, Conversion script, CouldBeWrong, Courcelles, Cplakidas, Crcd, Crowbarthe1337h4x0r, CryptoDerk, D, D.M. from Ukraine, DHN, DYLAN LENNON, DanMS, Dandrake, Daniel Quinlan, Danny, DarkFalls, Darkwind, Davewild, David Eppstein, David Pierce, DavidMCEddy, Davidiad, Dcarriso, Dekisugi, Delldot, Denny, Deodar, Deor, Deville, Dger, Dgw, Dick Shane, Diego Grez, Dimadick, Dimboukas, Diomedes1962, Discospinster, Dismas, Dlae, Dojarca, DominiqueM, DominvsVobiscvm, Doraemon168, DougsTech, Downtown dan seattle, Dp614912, Dr. Mriof, Dr.K., Dreadstar, DrunkenSmurf, Dumbo1, Dureo, Dwarf Kirlston, E. abu Filumena, EALacey, EL Willy, ENDWW, EauLibrarian, Eb.hoop, EdChem, EdJogg, EdwardLane, Egthegreat, Ehistory, Eisnel, El C, El Greco, El PDC, Elipongo, Eliz81, Elizabeyth, Emperorbma, Epbr123, Erkcan, Esperant, Espertus, Espin01, Espnguys, Eubulide, Euchiasmus, EugeneZelenko, Eukesh, Evandaeman, Everyking, Evil saltine, Ewlyahoocom, Ex0pos, Extransit, Fabrictramp, Falcorian, FeanorStar7, FellGleaming, Filthybutter, FlavrSavr, Flockmeal, Flyguy649, FocalPoint, FoekeNoppert, Fonzy, Footiemeister, Foozy101, Fordmadoxfraud, Fram, FranksValli, Fratrep, Frecklefoot, Fred64, Fredora, Fredrik, FreeKresge, Fullstop, FunPika, Func, Furrykef, Fxer, G.W., GPHemsley, Gaius Cornelius, Galoubet, Gamaboi, Gauge, Gcanyon, Gene Nygaard, Geneb1955, General Veers, Gfdsa, Ghepeu, Gidonshaviv, Giftlite, Gilliam, Gimboid13, Glenn, Godzillaspawn, Gogo Dodo, Goldfinch, GraemeL, Grafikm fr, Graham87, Grahamec, Great Deku Tree, GreekHouse, Gregbard, Grendelkhan, Groyolo, Grstain, Gun Powder Ma, Gurch, Guy Peters, Gwernol, HJ Mitchell, Hacker's union, Hadal, Hairy Dude, Harland1, Harrisjayjay, Harryboyles, Headbomb, Heimstern, Hemmingsen, HenryLi, Henrygb, Herbee, Hersfold, HiFlyer, Hullaballoo Wolfowitz, Husond, Hut 8.5, IHendry, Iamunknown, Ianmacm, Icairns, Ilario, Ilmari Karonen, Iluvswish, Imcoolerthanu, Imusade, Infrogmation, Inhumandecency, Intelligentsium, Inter, Intranetusa, Inwind, Ioeth, Iridescent, Ironholds, Isilanes, Islander, Ixfd64, J-stan, J.delanoy, JDPhD, JDoorjam, JIP, JRM, JSquish, JYolkowski, Jack100222, Jackerhack, Jagged 85, Jamaal224, James Balti, JamesAM, Jan eissfeldt, Jan.Kamenicek, Jaraalbe, Jaredwf, Java7837, JayJasper, Jbergquist, Jclerman, Jeff3000, Jesset77, Jgtl2, Jiang, Jim.belk, JimWae, Jitse Niesen, Jmlk17, Jnc, JoJan, JoanneB, JoeBlogsDord, Joelr31, John Reaves, John254, Johnbod, JohnnyCalifornia, Jojit fb, Jon Awbrey, Jona Lendering, Jonathunder, Jonel, Joowwww, Jorvik, Josh Grosse, Josh Parris, Jpk, Julia Rossi, Juliancolton, Jumbuck, Jumier, Junglecat, KRBN, Kaisershatner, Karanacs, Karl Andrews, Kasparov, Katefan0, Katimawan2005, Kauffner, Keegan, Keilana, Kenswarrior, Kenyon, Kewp, Khukri, Kimon, King of Hearts, KirbyRider, Kisonecat, Kkm010, Klhuillier, Koavf, Kukini, Kuru, LA2, LAX, LOTRrules, LWF, Lambiam, Laterensis, Lavo is straight, Ld100, Leranedo, Lesterjohn, Lighted Match, Lightmouse, Likebox, Linas, Llort, Lordvos, Lself, Lucidish, Lumpy27, Luna Santin, Lunkwill, Lupo, Lyndsaylove101, M a s, MC MasterChef, MC10, MER-C, MKar, MPerel, MPolaris, MZMcBride, Madmartigan1340, Magioladitis, Magister Mathematicae, Mailer diablo, Majorly, Malcolm Farmer, Mani1, MartinHarper, Mas 18 dl, Massivetigerdog, Materialscientist, Matia.gr, Mato, Matticus78, Maury Markowitz, Mav, Maxedesa, Mayooranathan, Mazca, Maziar.rezaei, Mcorazao, Mdd, Mdebets, Meekywiki, Meelar, Melchoir, Melesse, MeltBanana, Menthaxpiperita, Meteorit, Mhartl, Mic, Michael C Price, Michael Hardy, Midgrid, Miguel, Mike Christie, Milogardner, Mindmatrix, Mini-Geek, Minirogue, Mirwin, MisfitToys, Miskin, MisterSheik, Mitsuhirato, Mojska, Monobi, Montrealais, Moonriddengirl, Moreschi, Mpom13, Mr Stephen, Mr.Gnome, MrClock, Msanto, Muchclag, Mugunth Kumar, MusicStar201, Musunurijayachand, NHRHS2010, Naddy, Nakon, Nalco, Natalie Erin, NavarroJ, NawlinWiki, Ndavies2, NeuronExMachina, Nev1, Newkai, Newone, NickW557, Nicop (Usurp), Nietzsche1357, Nihiltres, Nine Tail Fox, Nishkid64, Nitya Dharma, Nixdorf, Nk, Nlu, No Guru, No1lakersfan, No31328, Noah Salzman, Noerdosnum, Nooristani, NuclearWarfare, ONUnicorn, Occono, OhanaUnited, Oleg Alexandrov, Olki, Omcnew, Omegatron, Omicronpersei8, Omnipaedista, One, Oolongy, Orioane, Out of Phase User, Owen214, Oxnard28, P0lyglut, Paine Ellsworth, PamD, Panarjedde, Parhamr, Paul A, Paul August, Pbroks13, Pedro, Pengo, PericlesofAthens, Persian Poet Gal, Peruvianllama, PeterSymonds, Peterlewis, Pewwer42, Pgecaj, Pgk, Philip Trueman, Physicists, Pie4all88, PierreAbbat, Piledhigheranddeeper, Pinethicket, Pjamescowie, Pjmc, Pmanderson, PoOpIeHeAd101, PoccilScript, Positron, Powerpeee, Prodego, Ptrillian, Quintote, Qxz, R'n'B, R.e.b., RDBury, RIrvine6204, RJHall, Radon210, Ragesoss, Rajah, Rakin295, Rambam420, RandomXYZb, Randomblue, Raoulis, Raul654, Readro, Reagle, Reaper X, RedMC, RedWolf, Redvers, Regentagger, Reid, Reisio, Remember, Remember the dot, Reverting, RexNL, Rhobite, Rholman8552, Rholton, Ricardobob, RickK, RjLesch, Rjwilmsi, Rk91, Rl, Rlevse, RobertG, Rocketman116, Ronark, Rotem Dan, Roundhouse0, RoyBoy, SEWilco, SWAdair, SallyForth123, Sam Hocevar, Sam Korn, SandyGeorgia, Sango123, Sannse, Sardanaphalus, Saruha, Scarian, Sceptic1954, Sceptre, Schkzamian, Scoo, Scorpion451, ScottW, Sdornan, Seanwhim, SebastianHelm, Segv11, Selfworm, Selket, Shanel, Shanes, Shaun ripcurl, Shedinja500, Shipmaster, Siddhant, SilverSurfer314, Simetrical, Sintonak.X, Slash, Sleeping123, SlimVirgin, Slushy9, Smack, Smashville, Smeira, Smyth, Snowolf, Sobaka, Soliloquial, Somno, SonOfMan, Sophia, Sopoforic, Spinorbit, Spiro Liacos, SpuriousQ, Squiddy, Srleffler, Steeev, Steel, Stefan Kühn, Stefanomione, SteinbDJ, Stephen MUFC, StephenBuxton, Stephenb, SteveLamacq43, Stevertigo, Stmoose, Storkk, StradivariusTV, Sub12, Subdolous, SugnuSicilianu, Sunderland06, Superbeecat, Superm401, Suruena, Susan Mason, Sushant gupta, Symane, Synthiac, THEN WHO WAS PHONE?, Tablizer, Tagishsimon, Tanvir Ahmmed, Tapir Terrific, Tbhotch, Tcncv, Teenagers4life, Tempshill, Terence, The ARK, The Singing Badger, The Thing That Should Not Be, The idiot, TheDJ, TheObtuseAngleOfDoom, Therefore, Thisisentirelyfactual, Thue, Tiggerjay, Timothy Clemans, Titanium Dragon, Titoxd, Tkuvho, Tmopkisn, Toh, Tom, Tom harrison, Tombomp, Tomruen, Tonyle, Tos, Tpbradbury, Trevor MacInnis, Tripbeetle, Triwbe, Twas Now, Twthmoses, Tzartzam, Ugen64, Uoregonduckman, UtherSRG, VASANTH S.N., VG Editor, Vacuum, Valery Beaud, Van helsing, Vanished User 0001, Venera 7, Vikaszt, Vilcxjo, Virginia-American, Viriditas, VivaEmilyDavies, Vladolena, Vybr8, WODUP, WWC, Wafulz, Wahabijaz, Wannbe Pink Bunny, Wareh, Wavehunter, Wenli, When1eight=2zeros, Wik, Wiki alf, WikiPhu, Wikianon, Wikieditor06, Wildhartlivie, Willa-a, Willseychew, Winhunter, Wizardman, WolfmanSF, WormRunner, Ww, Ww2censor, Wwoods, XJamRastafire, Xaosflux, Xharlequinxgrlx3, Xiner, Xmastershake, Xxlexixx, Yarnalgo, Yonatan, Zanibas, Zigger, Zundark, Ævar Arnfjörð Bjarmason, 'אנקדמנ, , 1441 anonymous edits

Diophantus *Source*: http://en.wikipedia.org/w/index.php?title=Diophantus *Contributors*: 3rdAlcove, 66.153.24.xxx, Aaron Rotenberg, Adt89, Afinebalance, Aitias, Alexius08, AllanBz, Andre Engels, Animum, AnonMoos, Anpersonalaccount, AppuruPan, Arcfrk, Arthena, Arthur Rubin, Asyndeton, Athenean, Awwiki, Barbatus, Bender235, Blanchardb, CRGreathouse, Can't sleep, clown will eat me, CanadianLinuxUser, Catalographer, Chaleyer61, Charles Matthews, Choochootrain1234, Chris55, Cje, ClanCC, Conversion script, CuteHappyBrute, D6, DDima, David Eppstein, Davidiad, Dawn Bard, Ddunx46135, Deeptrivia, Destroyer35, Dietary Fiber, Digitalmaster287, Dominus, EIF1, EauLibrarian, Empire World Champ, EoGuy, Erud, Everyking, Excirial, Falcon8765, Farosdaughter, Fawcett5, FeanorStar7, Fenners, Flamingspinach, Floquenbeam, Gabbe, Gamonetus, Gauss, Georg Muntingh, Giftlite, Grafen, Gun Powder Ma, Gurch, Gwernol, Haham hanuka, Hardmath, Hede2000, Hobojoe9127, Icairns, Inmaculada Perez, IronChris, Jack Phoenix, Jagged 85, JamesMLane, Jaredwf, Jim1138, Jpvandijk, Knight1993, Kylu, Lambiam, LeeG, Lipoff, Llywrch, Luk, MONGO, Marc Venot, Margacst, Markdraper, Marshman, Mcorazao, Michael C Price, Mindmatrix, Miskin, Mukadderat, Musunurijayachand, Natalie Erin, Neil Parker, Nicolas Perrault III, Nikbakht21, Numbo3, Omeganian, Ontoraul, Opoudjis, Oxymoron83, Pbroks13, PerezTerron, PericlesofAthens, Peruvianllama, Petrouchka, Philip Trueman, Piano non troppo, Psy guy, QuadrivialMind, R'n'B, Ratfox, Raven in Orbit, Reedy, Rich Farmbrough, RisingSunWiki, Ronhjones, Runescape-rares admin, Ruud Koot, Ruwanraj, S lijin, S711, SMasters, Salgueiro, Salix alba, Schutz, Seaphoto, Selfworm, Siyaz, Skizzik, Somno, Staka, Stevertigo, StradivariusTV, TKD, Tankred, Tgeairn, The Thing That Should Not Be, Thingg, Tiddly Tom, Tkuvho, Tox, Trovatore, Ttiotsw, Tzartzam, Velella, Versus22, VivaEmilyDavies, Wareh, Wikipelli, William M. Connolley, Wmahan, XJamRastafire, Yamamoto Ichiro, Zahid Abdassabur, Zepheus, وخور ,يمانا, 314 anonymous edits

Euclid *Source*: http://en.wikipedia.org/w/index.php?title=Euclid *Contributors*: (jarbarf), 10metreh, 1297, 15lsoucy, 3rdAlcove, 4twenty42o, 5 albert square, 62.202.85.xxx, A F K When Needed, A Train, A. Parrot, Abi79, Academic Challenger, Acroterion, Adashiel, Addshore, AdjustShift, Aftermath, AgadaUrbanit, Ahmedettaf, Ahoerstemeier, Aim Here, Airplaneman, Aitias, Alansohn, AlexiusHoratius, Ali, Allen234, AlotToLearn, Alpha 4615, Alsandro, Altenmann, Alxeedo, Amberrock, AnakngAraw, Andre Engels, Andrew Rule, AndriuZ, AndySimpson, AngelOfSadness, Animum, AnonMoos, Anonymous Dissident, Anonymous101, Antandrus, Ante Aikio, Anu cool hot, Apbennett, Apparition11, Arcfrk, Arichnad, Arthena, AstroNox, Astrochemist, Asyndeton, Athenean, AuburnPilot, Autoerrant, Avicennasis, Avs5221, AxelBoldt, B0OP, B4hand, BD2412, Badgettrg, Bakilas, Bamfrofl, Banus, Bcrowell, Beaunickerson, Beetlegeuise, Bejesus, Ben Jos, Bharatveer, BillLoney, Binabik80, Binksternet, Blanchardb, Blueboy814, Bobo192, Bobrayner, Bongwarrior, Bookandcoffee, Borisblue, Brian Pearson, Brion VIBBER, Brookie, Brutannica, Bunnyy67, Bunzil, C0N6R355, CH3rryCHiiCK, CHITRANI, CMacMillan, Cactus.man, Cake taken, Callummendoza, Caltas, Can't sleep, clown will eat me, CanadianLinuxUser, Cantiorix, Capricorn42, Captain Curry, CardinalDan, Carmichael, Cartiod, Catalographer, Caulde, Cflm001, CharlotteWebb, Chepner, Chochopk, Chris19910, Christian List, Cje, Ckatz, Clovis Sangrail, Comancheros, Comfortably Paranoid, CompRhetoric, Connormah, Contribution, Conversion script, Corpx, Courcelles, Croperz, Cst17, Ctbolt, DARTH SIDIOUS 2, DBaba, DGJM, DNewhall, DVD R W, Dacker13, DanMS, Daniel5127, DanielCD, Danjel, Dannychaseismyname, Dantheman531, Darren23, Dart169, Darth Panda, David0811, Dcljr, DeadlyHobo, Deeahblita, Defender of torch, Deflective, Delldot, Den fjättrade ankan, DerHexer, Deutschgirl, Devbhargava, Dgies, Dickdock, Diemunkiesdie, Dimboukas, Dina, Dino, DirectEdge, Discospinster, Ditcuc, Dmitri Yuriev, Dogru144, DolphinL, Don4of4, Donthedon8675, Download, Drift chambers, Drmjam, Drumguy8800, DuKu, Dudepizza19, EJF, ESkog, EauLibrarian, Edward, Egthegreat, El C, Elderleo, Eleuther, Elmerson, EmilJ, Empty Buffer, Enviroboy, Epbr123, Epolk, Eric119, Eubulides, Euryalus, Excirial, Extransit, Fang Aili, Farshadrbn, Favonian, FayssalF, FeanorStar7, Fieldday-sunday, Finell, Fjbfour, Flyguy649, FollowTheMedia, FrancisTyers, FreplySpang, Frood, Funeral, Fyyer, Fæ, GTBacchus, Gadfium, Galoubet, Gampy, Gary King, Gatherton, Gawaxay, Gene.arboit, Ghaly, GhostPirate, Giftlite, Giggy, Gigogag, Gildenlow, Gilliam, Ginkgo100, Ginsengbomb, Goken13, GoonerDP, GorillaWarfare, Goudzovski, GraemeL, Graham87, Gravitan, Gtg204y, Gurch, HJ Mitchell, Haaqfun, Hacky dude93, HaeB, Hakkinen, Hargrimm, Haukurth, Helopticor, Hippophaë, HoodedMan, Hut 8.5, Hyacinth, Hypergeek14, I dream of horses, Icairns, Icestorm815, Ideyal, Igoldste, Imminentnematodes, Immunize, Imroy, Instinct, Integer Man, Io anthony, Iridescent, Irish Souffle, It Is Me Here, Ixfd64, J.delanoy, J8079s, JFreeman, Ja 62, Jagged 85, Jake Wartenberg, JamesBWatson, JamieS93, Jane Bennet, Jaredwf, Jayjaypieman, Jebus989, Jeffrey Mall, Jengod, Jennavecia, JesseGarrett, Jfdwolff, Jgtl2, Jhtay, Jiddisch, JimWae, Jj137, Jncraton, Jni, Joevandy, John Hyams, John.d.page, JohnCD, Johnbibby, Johnskate17, Jondel, Jonomacdrones, Joshuabowman, Jossi, Jpgordon, Jsharpminor, Juliancolton, Jusdafax, Jyusin, KGasso, Kacembepower, KamuiShirou, Kazvorpal, Kcordina, Keegan, Keilana, Kf4bdy, Kibibu, Killiondude, Kku, Klow, Kocio, Kribbeh, Krich, Krun, Kruusamägi, Kubigula, KudulO, Kuru, Kurykh, Kwamikagami, L Kensington, La comadreja, Lacatosias, Lachlan wilson, Lapisphil, Latka, LeaveSleaves, Ledzep123strat, Lerdthenerd, Leszek Jańczuk, Leutha, LilWille, Little Mountain 5, LittleLottie, LittleOldMe, LizardJr8, Loadmaster, Lolcats121567, Looxix, Loren.wilton, Lotje, Lova Falk, Love And Fellowship, Luke.haney, Lycurgus, MER-C, MPerel, Macedonian, Madhero88, Mahmudmasri, Malcolmxl5, Malhonen, Malo, Mandarax, Marc Venot, Marek69, Mark Wolfe, MarkGallagher, Markjoseph125, Markus Poessel, Martin451, Mathewsyriac, Mato, Matthew Fennell, Maxim, McSly, Meaghan, Melsaran, MerchantOfBaggage, Message From Xenu, Michael Hardy, MichaelTinkler, Miciah, Mike Rosoft, Mikeo, Mingshey, Mini-Geek, Miskin, Mjr162006, Mmoneypenny, Mnemosyne89, Momo san, Monty845, Moreschi, Mormon17, Moster01, Mousty6, Mr. Lefty, MrJones, Mrgregg108, Mu301, MultiFroggie1, Munita Prasad, Munstaslaya6, Mygerardromance, N5iln, NHRHS2010, Nakon, NawlinWiki, Nehrams2020, NeighborhoodGenie, Nekura, NeoNerd, Nev1, NikolaiLobachevsky, Nilfanion, NittyG, Nixdorf, Njaikrishna, Nlu, Norm mit, Novangelis, NuclearWarfare, Numbo3, Ocaasi, Octoberer, Oddity-, Olivier, OllieFury, Omicronpersei8, Omnipaedista, OneWeirdDude, Optichan, OverlordQ, OwenKL, OwenX, Oxymoron83, Pagrashtak, Panarjedde, Patstuart, Paul August, Pb30, Pbroks13, Pegasus1138, Peruvianllama, Petrouchka, Phantomsteve, Pharaoh of the Wizards, Philip Trueman, Piano non troppo, Pinethicket, Pinkicc65, Pjbrooklyn, Ponyo, Poppi12, Prb4, PrestonH, Prodego, Prolog, Pupster21, Pyroclastic, Python eggs, Quanticles, Quintote, Qxz, RDBury, RJaguar3, Ractogon, Rafaeltim, Rasenshuriken, Rawrcopter, RayKiddy, Recurring dreams, Rich Farmbrough, Richard B, Risk one, Rj, RobertG, Robma, Ross Burgess, Rothorpe, RoySmith, Ryan Reich, Salix alba, Sam Blacketer, Sam Hocevar, SamuelTheGhost, SarekOfVulcan, Sbrools, Scarian, Scientific29, Scootey, Scriberius, Sebastian Goll, Selfworm, Shadowjams, Shanes, Shipmaster, Shisir 1945, Shizhao, Shlomke, Shotmenot, SimDarthMaul, Sitethief, Skokos, Skomorokh, Smartman789, Smeira, Smsarmad, Snowolf, Someguy1221, Soumyasch, SpaceFlight89, SpeedyGonsales, Spireguy, SpookyMulder, Squids and Chips, StNicksRocks, Staffwaterboy, Starry maiden Gazer, Stbalbach, Stephenb, SteveCaton, Stevertigo, Storm Rider, Stradv16, Stubblyhead, StupidYellow14, Sumahoy, Sunray, SuperGirl, Sven Manguard, Symane, Syncategoremata, Sławomir Biały, TEB728, THEN WHO WAS PHONE?, Tangotango, Tanthalas39, Tanvir Ahmmed, Tarret, Tbhotch, Tchoř, Tcncv, Tempodivalse, Terrx, TeshawnEdmonds, ThaddeusB, That Guy, From That Show!, Thatguyflint, The Thing That Should Not Be, TheXenocide, Thecheesykid, Theda, Thehelpfulone, Thrind, Thw1309, Tide rolls, Tinlv7, Tiptoety, Tom harrison, Tommy2010, Tpbradbury, Trashmetal135, Traxs7, Trevor MacInnis, Triwbe, Tsp greenday, Twas Now, TylerJarHead, Tyoung3, UDScott, Ukexpat, Ulner, Unschool, Useight, Utcursch, VI, VSimonian, Vahagn Petrosyan, Vary, Vaughan Pratt, Vaxelrod, Vector Potential, Veghead, Venera 7, Viktor Laszlo, Vivio Testarossa, VolatileChemical, Vrenator, Vsmith, Wafulz, WatermelonPotion, Wavemaster447, Wayne Slam, Weston.pace, Wetman, Whitemastiff, Whouk, Wiki alf, WikiLaurent, Wikiklrsc, WikipedianMarlith, Wikipelli, Will2710, Willking1979, Wilson44691, Wimt, Wireless Keyboard, Witbyt, Wmahan, Ww, XJamRastafire, Xelgen, Xx3nvyxx, Yamamoto Ichiro, Youssefsan, Yuval Madar, Zirland, Zootm, Zundark, Саша Стефановић, , 1750 anonymous edits

Eudoxus_of_Cnidus *Source*: http://en.wikipedia.org/w/index.php?title=Eudoxus_of_Cnidus *Contributors*: .:Ajvol:., AK456, Abogom, Adashiel, Albert Non, Aldoaldoz, Alethe, Altairisfar, Alyssa3467, AnnaP, Arthena, Atchom, Avatar, Awolf002, B210303V3941W42371, Bender235, BertSen, Binabik80, Brona, Caltas, Can't sleep, clown will eat me, CanadianLinuxUser, Catalographer, Cessator, Charles Matthews, ChrisGualtieri, Chronicler, Constructive editor, Ct74848, Curps, Cwkmail, DMacks, Dandrake, Dawn Bard, DeadEyeArrow, Dentalplanlisa, Der Spion, Dimadick, Dissolve, Dominus, Dr. Submillimeter, Drdonzi, Dycedarg, ELApro, EauLibrarian, Eclecticology, Eroica, FeanorStar7, Gauss, Gene Ward Smith, Giftlite, Haiduc, Icairns, Igiffin, IronChris, JBdV, Jaraalbe, Jaredwf, Jmlk17, John Gibbons 3, Juliancolton, KRBN, Kazvorpal, Koranjem, Kostisl, Kubigula, Kuru, Lampica, Leeiscool56, Lzur, MER-C, Machine Elf 1735, Maestlin, Magnus Manske, Mandarax, MarcoTolo, Maurog, Michael Hardy, Miguel, Mikemoral, Moorlock, Msanta800, Newbyguesses, Nicke Lilltroll, Nk, Novangelis, Olivier, Omnipedian, Oxymoron83, Ozob, Peruvianllama, Philip Trueman, Phædrus, Plasynins, Pollinosisss, Polylerus, Pranathi, R'n'B, Rich Farmbrough, SDC, Selfworm, Silly rabbit, Singinglemon, Slakr, Smokizzy, Some jerk on the Internet, Soulfare, Spacepotato, Stevertigo, Tassedethe, Tom Peters, Tomaxer, Tothebarricades.tk, Wmahan, XJamRastafire, YUL89YYZ, Zbayz, Zeeyanketu, תמאה תויסחי, 111 anonymous edits

Hipparchus *Source*: http://en.wikipedia.org/w/index.php?title=Hipparchus *Contributors*: 1c33y37, 1exec1, 200.191.188.xxx, 2001:558:6026:81:7560:20C2:D985:2A70, Ahoerstemeier, Alaric evans, Aldux, Andre Engels, Anna Lincoln, Anonymous Dissident, Antandrus, Anthony, AstroNomer, AxelBoldt, Az1568, Aznroyal, Bagatelle, Bejnar, Belovedfreak, Bensmith53, Biztalkguy, Blaxthos, Braincricket, Bryan Derksen, Burek, CDN99, CMPtnc, CTZMSC3, Calor, CanisRufus, Captain Segfault, Catalographer, CharlotteWebb, ChristopherWillis, Clarince63, Claygate, Closedmouth, Cobness, Codex Sinaiticus, Connection, Conversion script, Coren, Cplakidas, Crculver, Curps, Cyberstrike3000X, Dantheox, Darlene4, David Eppstein, Davidiad, Dennis Valeev, Dimadick, Dirtyasianmen, Djnjwd, Dom497, Donjanssen, Dougweller, Dr. Submillimeter, Dryoldscholar, ELApro, EauLibrarian, Edcolins, Edwinstearns, EmilJ, EricWesBrown, Eumolpo, Excirial, Favonian, FeanorStar7, FisherQueen, FocalPoint, Fotaun, Francis Schonken, FrancoGG, Francs2000, Frehley, Gdr, Geboy, Gene Nygaard, Ghane, Giftlite, Gilgamesh, Gilliam, Goretsky, GorillaWarfare, Graham87, Gun Powder Ma, Gurch, Hadal, Hajor, Hede2000, Hephaestos, Heron, Hmains, Honbicot, Hopa220, Huey45, I dream of horses, Icairns, Immunize, Inferno, Lord of Penguins, JMK, Jaberwocky6669, Jagged 85, Jaraalbe, Jaredwf, JayJasper, Jennavecia, Jhnmurphy, Jigj119, Jimp, Jj137, Jkelly, Jmccann, Joe Kress, Johntex, Josh Grosse, Juicyschnapper, Jyril, Karl Palmen, Keithpickering, Ken Drummond, Kepler60, Kevin chen2003, Kilcoyne, Kl4m, Korean alpha for knowledge, L Kensington, LAAFan, Latiniensis, Leal Nightrunner, LeaveSleaves, Leithp, Les Meloures, Lissajous, Little Savage, Looxix, Lotje, Lupin, Lzur, Macedonian, Machine Elf 1735, Maestlin, Martarius, Maurog, Mdebets, Mh26, Michael Daly, Michael Hardy, Michael Snow, MichaelTinkler, Mike s, Minesweeper, Miracle Pen, Mishuletz, Mkl12, Mlm42, Mlpearc Public, Mmm, Modeha, Modest Genius, Montrealais, Moonsword, MrDolomite, NHRHS2010, NawlinWiki, Noctibus, Notesenses, Numbo3, Ordinaterr, Owl 2, P0per, Panairjdde, Panarjedde, Paoai, Parvathisri, Paste, Paul August, Paul Drye, PericlesofAthens, Persian Poet Gal, Peruvianllama, Pezanos, Philip Trueman, PierreAbbat, Pinethicket, Piolinfax, Pizza Puzzle, PlanetStar, Plasynins, Pmanderson, Possum, Praefectorian, Praveenp, Pro crast in a tor, Professor marginalia, Puny1234, Pustelnik, Pylambert, Python eggs, RJHall, Ragesoss, RainbowOfLight, Reach Out to the Truth, Rich Farmbrough, Roymail, Rursus, Rwflammang, SG Slugger, Sabri76, Salix alba, Sam Hocevar, Sburke, Sceptic1954, Schneedog1, Sdfg, Searchme, Segv11, Selfworm, Seraphimblade, Serasuna, Shantavira, Shizhao, Silver7scythe7, Singinglemon, Skamecrazy123, Skeptic2, Skwidz, Snowolf, Snoyes, Soliloquial, Stan Shebs, Staszek Lem, Stevenmitchell, Supersammyboy, Syncategoremata, Tabletop, Tercer, Terence, Terry0051, Terrymacro, The Man in Question, Thorwald, Tide rolls, Tom Peters, Twang, Unyoyega, Urhixidur, UtherSRG, Vgent, Viriditas, Wareh, Welsh, Wik, WinoWeritas, XJamRastafire, Zundark, 340 anonymous edits

Hero_of_Alexandria *Source*: http://en.wikipedia.org/w/index.php?title=Hero_of_Alexandria *Contributors*: 1brettsnyder, 2D, 3rdAlcove, Academic Challenger, Ahoerstemeier, Amux, Anakata, AnakngAraw, AndreasJS, Andymeyer, Arthena, AxelBoldt, Axeman89, Bachame, Bensmith53, Biscuittin, Bkwillwm, Blaxthos, Bob Burkhardt, Bogdangiusca, Bondaruk85, CWii, Carjoyg, Catalographer, Chaocoine, Charles Matthews, CharlesGillingham, Charlesdrakew, Chetvorno, Chibi5100171, Chris the speller, ChrisGualtieri, Christian List, Ciphers, Cjassebi, CommonsDelinker, Cool Blue, Corpx, CuteHappyBrute, Dan6hell66, Dante Alighieri, David1955, DeadEyeArrow, Denseboy, Der Golem, Dfrankow, Diverman, Dj245, Djnjwd, Djrdjn, DopefishJustin, Dougk, Dougweller, Downwards, Dungodung, Dunnhaupt, Dureo, EJF, EauLibrarian, EdJogg, Edwinstearns, Egil, Enquire, Esn, Feezo, Finn-Zoltan, Fjbfour, Foxcreeper, FranksValli, Frotz, Gaiusknight, GalanM, Gdr, Germandemat, Giftlite, Gilliam, Giorgos Tzimas, Goethean, Graham87, Granite07, Gtg204y, Gun Powder Ma, Gurps npc, Hanacy, Heerok.banerjee, HenkvD, Herbee, Heron, Hockey2009, Hornlitz, Huey45, Hugo999, II MusLiM HyBRiD II, Ian Pitchford, Iceberg3k, Igiffin, Immunize, Infrogmation, Ipatrol, J.delanoy, J8079s, JHunterJ, Jagged 85, Jaraalbe, Jbergquist, Jeffrey O. Gustafson, Jgtl2, Joe Dick, John, John-1107, Johncatsoulis, Jonas Kölker, Josh Parris, Jpgordon, Juanc.suar, Kairos, Katalaveno, Kelson, Khoikhoi, Kiko4564, Kilik Kinak34, KnowledgeOfSelf, Lackthereof, Lenzorg, Light current, Lishuguy, Lisiate, Lysdexia, Lzur, MJ94, Macedonian, Macrakis, Magnus.de, Makemi, Marcuswaterman, Marudubshinki, Mathew5000, Matia.gr, Message From Xenu, Micahbode, Michael Hardy, Michaelbusch, Minesweeper, Miro1040, Miskin, Miwanya, Mlessard, Modalfinatto, Mononoke, Mschlindwein, Myfanwy, Narwhal2, NerdyScienceDude, Newton2, Nk, NotAnonymous0, Nousernamesleft, O.Koslowski, Old Moonraker, Oleg Alexandrov, P.B. Pilhet, Paleorthid, Paul A, PericlesofAthens, Persian Poet Gal, Peterhi, Peterlewis, Petrouchka, Philip Trueman, Pixel ;-), Pogonatos2, Punditah, Quadell, Qxz, R'n'B, Raymondwinn, Rich Farmbrough, Richard Arthur Norton (1958-), Rjwilmsi, Sailsbystars, Scarian, Scorpion451, Selfworm, Shirulashem, Siddhant, Slakr, Sm53629, Soliloquial, SpaceFlight89, Spamhunt, Srleffler, Steve10345, Steven Zhang, Subversive.sound, Swimaholic623, Syncategoremata, TOO, The way, the truth, and the light, TheOldJacobite, Thetorpedodog, Thumperward, Tide rolls, Tintinlover123, Twang, TwistOfCain, Vald, Vary, Viriditas, VivaEmilyDavies, WAS 4.250, Wizard191, Wknight94, Wolfkeeper, Woohookitty, XJamRastafire, Xiphoris, Yintan, Zachariel, Zenyu, זיו, 374 anonymous edits

Nicomachus *Source*: http://en.wikipedia.org/w/index.php?title=Nicomachus *Contributors*: Achowat, AllanBz, Arcadian, BMBTHC, Brando130, Catalographer, Chuckiesdad, Ernobe, FeanorStar7, Fraytel, Gdr, Gene.arboit, Giftlite, GreatWhiteNortherner, JYOuyang, JamesAM, Jaraalbe, Jaredwf, Jbergquist, Jgtl2, Jim McKeeth, Jossi, KRBN, Lykos, Maymay 1699, Mscuthbert, Neilc, Olaf, Pasicles, Pezanos, RekishiEJ, Rgvedder, Sannse, Selfworm, Shadowjams, Singinglemon, Soufray, Stevenmitchell, Sullivan.t.j, Swamp Ig, Tamfang, TheOldJacobite,

Tomisti, Tony Ant, WVhybrid, Wareh, WhiteC, XJamRastafire, 21 anonymous edits

Menelaus_of_Alexandria *Source*: http://en.wikipedia.org/w/index.php?title=Menelaus_of_Alexandria *Contributors*: 3rdAlcove, Abrech, Agutie, Anlace, Antonio Lopez, Big Bob the Finder, Catalographer, Conone, CuteHappyBrute, Daniel Mahu, Davidallarddotcom, Devin, FeanorStar7, Fedallah, Florian Huber, Hello71, Hégésippe Cormier, Icairns, Igiffin, Intelligentsium, Jagged 85, Jaraalbe, Jaredwf, Jose Ramos, Maestlin, Margacst, Maurog, MichaelTinkler, MrOllie, PasabaPorAqui, Petrouchka, Phe, Polylerus, Promethean, RDBury, Sam Blacketer, Scetoaux, Selfworm, Singinglemon, Skeptic2, Surya31415, TOO, TheGrappler, Tokek, Uriber, WojPob, Wtmitchell, XJamRastafire, Zad68, Zaslav, Zerida, Zzuuzz, 23 anonymous edits

Pappus_of_Alexandria *Source*: http://en.wikipedia.org/w/index.php?title=Pappus_of_Alexandria *Contributors*: 3rdAlcove, AllanBz, Arcfrk, Arthena, Athenean, Beetlegeuise, Catalographer, Charles Matthews, Cje, CuteHappyBrute, Danny, Dantheox, Davidiad, Deagle AP, Delirium, Deucalionite, Dickdock, Dino, Dogru144, Dungodung, ESkog, Ejosse1, EmilJ, FeanorStar7, FocalPoint, Giftlite, GoingBatty, Icairns, Igiffin, Ishi Gustaedr, Jagged 85, Jaraalbe, Jaredwf, Jim.belk, Joyous!, Jwoodger, KRBN, LeaveSleaves, Lestrade, Lzur, Maestlin, Michael Hardy, Nic bor, Ojigiri, Oktanyum, Parneix, Pbroks13, Petrouchka, Polylerus, Quiet Editor, RDBury, RJFJR, Rgdboer, Robertsoncrusoe, RustySpear, SMasters, Salix alba, Sburke, Scorwin, Selfworm, Sumahoy, SusikMkr, TakuyaMurata, Tokek, Tonalone, VivaEmilyDavies, Vsmith, Wareh, XJamRastafire, Zerida, 65 anonymous edits

Pythagoras *Source*: http://en.wikipedia.org/w/index.php?title=Pythagoras *Contributors*: 19.168, 3rdAlcove, 96.186, A bit iffy, A. di M., A.J.Chesswas, AAA!, ACupOfCoffee, APMiller, Abach, Academic Challenger, Achilles, Across.The.Synapse, Acroterion, Adam Bishop, Adam Conover, Addshore, Adriaan, Adrian Sampson, Adzscott, Aelffin, Aeonx, Aeternus, Aetheling, Aforencich, AgentFade2Black, AgentPeppermint, Agricultor, Ahoerstemeier, Aitias, Ajraddatz, Aksi great, Ale jrb, Aleenf1, AlexDenney, Algorithme, Alphachimp, Altenmann, Alynna Kasmira, Amberrock, Anandbrar, Andargor, Andhika93, Andkaha, Andrasnm, Andre Engels, Andres, Andrew Gray, Andrewpmk, Angrytaco123, Angusmclellan, Animum, Ann Stouter, Annielogue, AnonMoos, Anonymous Dissident, Anonymous anonymous, Anonymous56789, Antandrus, Antireconciler, Aprogressivist, Aquarius Rising, Archer7, Argos'Dad, Aristophanes68, Arjun01, ArmadniGeneral, Artaxiad, ArthurDuhurst, Arty 1999, Asportspro, Astrologist, Asyndeton, Athkalani, Atlant, August, AuntieMormom, AvicAWB, Avicennasis, Awh, Ayudante, AzaToth, B-Machine, B.d.mills, BD2412, BMW 720i, BVegan2, Bacchiad, Badagnani, Banes, Banpei, Barneca, Beckylou23, Beezhive, Belovedfreak, Ben davison, Ben-Zin, BenKovitz, BenTels, Bender235, Benevolentkisses, Bgold4, Bgs022, Bhugh, BigMacman, Bilbobee, BillShurts, Binabik80, Bioinformin, Blablaer, Blair Bonnett, Blockinblox, Blue520, Blueanode, Bob burkinson, BobTheTomato, Bobo192, Boboroshi, Bongwarrior, Bookworm857158367, Borgx, Bovlb, Brad Watson, Miami, Brazucs, BrianGV, Brion VIBBER, BrokenSphere, Bruce1ee, Brusegadi, Bsherr, Bubba hotep, Buickid, BullRangifer, Bumbumman, Bushytails, CBDunkerson, CJLL Wright, CRGreathouse, Cactus.man, Caesarjbsquitti, Caltas, Calvin 1998, CambridgeBayWeather, Can't sleep, clown will eat me, CanadianCaesar, CanadianLinuxUser, CardinalDan, Carpenoctem, Carrionluggage, Carton37172, Casius, Catalographer, Catalyst2007, Cavenba, Cenarium, Centaurioid, Ceoil, Chad1m, Chairboy, Chameleon, Charles Matthews, CharlotteWebb, CheeseDog, Chika11, Chill doubt, Chitomcgee, Chodorkovskiy, Chris 73, Chrislk02, Chronicler, Ck lostsword, Coelacan, Collins.mc, Come to paddy, Cometstyles, Connection, Connormah, Contest, Conversion script, Corpx, Crakinmackin, Curps, Cxz111, CzarB, D.Balaski, D.M. from Ukraine, DARTH SIDIOUS 2, DBaba, DIEGO RICARDO PEREIRA, DVD R W, DVdm, Da stooges, DanielCD, Danny-w, Danski14, DarkAvZ, Darkwind, Darrendeng, Dast, Davewild, David Underdown, Davidiad, Dawsonhenshaw, Dbachmann, Dddoooggg, DeadEyeArrow, Deadlyzen, DearPrudence, Deathgun456, Dedoleo, Dekisugi, Delbert Grady, Delta Tango, Den fjättrade ankan, DennyColt, Deor, DerHexer, Deville, Dick Emery, Dicklyon, Diliff, Dimadick, DiscordantNote, Dismas, Dissident, Dittaeva, Djproject, Dlohcierekim's sock, Dogwhelk, DonDiego, Donarreiskoffer, Donbert, Doug Bell, Dr spork, Dragomiloff, Dreadstar, Drg55, Drift chambers, Dualphasesaber, Dud94, Dycedarg, EALacey, ERcheck, ESkog, Eaolson, Earthwormdiscples, East718, Eatysoony, EdGl, Edgar181, Editmaniac, Editor2020, Edward321, Eequor, Ehman, Ej, Ekrub-ntyh, El C, Elbarto92, Emceerival, Emre D., Englishnerd, Eric Blatant, Eric Kvaalen, Eriscool 12, EuPhyte, EvanProdromou, Everyking, Evil Monkey, Exert, Extransit, FDR, FJPB, FRS, Fabrictramp, Fang Aili, Fastfission, FeanorStar7, Firetrap9254, FitzColinGerald, Flauto Dolce, Flowerparty, Flowerpotman, Fluri, Flying Hamster, Fordmadoxfraud, FranksValli, Frazzydee, Fredrik, FreplySpang, Fuhghettaboutit, Func, Furkhaocean, Futurebird, FvdP, Gabbo555, Gaff, Ganesh Paudel, Gary King, Gaudio, Gauss, Gekritzl, Geneb1955, General Wesc, Geni, Geniac, Gfoley4, Giftlite, Gilliam, Girlsrul101, Gitartha.bordoloi, Givepain, Glacier Wolf, Glen, Graham87, Greece666, GringoInChile, Grunt, Gsandi, Gscshoyru, Gurch, Guvnikov, Guyanesebaby, Gwernol, HJ Mitchell, HJK6, Hack-Man, Hadal, Hammer1980, Hanzaruhacheck, Harej, Hatmatbbat10, Hayabusa future, Hayjohn, Haza-w, Hazard-SJ, Herk1955, Hhhhhhj, HiDrNick, Hinotori, HisSpaceResearch, Hmrox, Honbicot, Hpc, Hu12, Hut 8.5, I already forgot, Ian.thomson, Iannigb, Icairns, Idiotoff, Igorwindsor, Ijhbdfvisrn, Ikiroid, Imasleepviking, Imusade, In Medias Spes, InfinityAndBeyond, Information Habitat, Infosocialist, Inwind, Irishguy, Isis07, Ixfd64, J.delanoy, J8079s, JDPhD, JDspeeder1, JForget, JFreeman, JSpung, JYolkowski, Jagged 85, James086, JamesR, Jamespeterka, Jami6361, Jamsponge, Janderk, Jaraalbe, Jaredwf, Jaxl, Jaybriggs, Jayjg, Jaymax, Jbamb, Jbergquist, Jbussell, Jclark256, Jcw69, JedStates, Jeepday, Jeff G., Jeffrey t b, Jennneal1313, JeremyA, Jfpierce, Jfsnjax, Jh51681, Jimokay, Jjjjjh, Jmlk17, JoanneB, Joe056, JohnSawyer, Johnleemk, Johnny Sumner, Jol123, Jon Awbrey, Jona Lendering, Jonathunder, Jonhays0, Jorf, Josh Parris, Jossi, JudiciousH, Jumbuck, Justanother, Justin Eiler, Jvbishop, Jóna Þórunn, K2111ATIE, Kaicarver, Kaisershatner, Kalki, Kanġi Oh̄anko, Karisade, Katalaveno, Katieh5584, Kbdank71, Kbh3rd, Kd050896, Kedi the tramp, Keilana, Keith Edkins, Kenders92, Kerotan, Kesiyaj, Ketsuekigata, Khoikhoi, Kimon, King of Hearts (old account 2), Kingkaos0, Kipala, Kirrages, Kkm010, Knakts, KnightLago, Knights Templar, KnowledgeOfSelf, Knucmo2, Koranjem, KoshVorlon, Kotuku33, Kptracey, Kribbeh, Krscal, Kubigula, Kukini, Kungfuadam, Kungming2, Kuru, Kwertii, Kyle Barbour, Kyriakos, Kzollman, L Kensington, LAX, LOL, La Pizza11, Lambiam, Laualoha, Leafyplant, Lebroyl, LeighvsOptimvsMaximvs, Leoadec, Leonidas Metello, Lights, LilWille, Lilac Soul, Linkact12, LiquidLife, LittleDan, Llywrch, Loco4240, Logannoonan, Logical2u, Longhair, Loren.wilton, Lorynote, LoveMonkey, Lucidish, Luk, Luna Santin, Lupo, M neimeyer, MER-C, MParshall, MPerel, Macintosh User, Magioladitis, Magister Mathematicae, Magnus Manske, Mahlum, Majorly, Mal4mac, Malaiya, Malcolm Farmer, Male1979, Mallanox, Mana2432, Mani1, Marc Venot, Marco Guzman, Jr, Mark Christensen, Marknau, Matt me, Mattdm, Matthewbc02, Matthuxtable, Maurice Carbonaro, Mav, Maverick423, Mayooranathan, Mbertsch, Me wolf, Meggar, Mentifisto, Merope, Metron4, Mets501, Meursault2004, Mgales, Mhking, Michael Hardy, MichaelBillington, Michaelas10, Michkalas, Mihai, Mike Segal, Mike gg5, Minna Sora no Shita, Mladifilozof, Mod.torrentrealm, Mordicus Egg, Mormonchess, Mr magnolias, Mr. Billion, Mr. Lefty, MrFish, Mrmoocow, Mscuthbert, Mungo Kitsch, My name, Myanw, NBeale, NHRHS2010, NKSCF, Nakon, Nareek, Nasz, Nava 31, NawlinWiki, Ndaco, Ndn, Ndokos, Netkinetic, Neverquick, Nhatsux4eva, Nick, Nick C, Nickeleon, NightMonkey, NikolaiLobachevsky, Nikolaos Bakalis, Nixdorf, Nlu, No Guru, Node ue, Norwikian, Nufy8, ObeyMe, Ocanter, Oda Mari, Odysses, Ogress, Oleg Alexandrov, Omicronpersei8, Onef9day, Optim, Orbst, Orecalimo, Otojin, OverlordQ, OwenX, Oxymoron83, PGWG, Pacific PanDeist, Pak21, Papa November, PaperTruths, Parsifal, Part Deux, Pascal.Tesson, Pasicles, PatrickA, Paul August, Pax:Vobiscum, Pbroks13, Pbrook, Pedant17, PeteShanosky, Peterlewis, Pezanos, Phatboi0.0, Philip Trueman, PhilipDSullivan, Philsopher Ben, Pigman, Pilotguy, Pinethicket, Piniricc65, Pizza Puzzle, Pjb3293, Plattler01, Polenth, Pollinosisss, Prb4, PrestonH, Prestonmag, Prodego, Pseudomonas, Punkasl3, PuppyLover2410, Quendus, Quinsareth, Quintote, Qwfp, Qxz, QzarBaron, R'n'B, R. S. Shaw, R.E. Freak, RDF, RJR3333, Radgeek, Randomjohn, Rashid 11, RazorICE, Rebecca Fleisher, Red Director, Redheylin, Rere65, Rettetast, Revolución, RexNL, Riana, Rich Farmbrough, Richard D. LeCour, Rick Norwood, Rjwilmsi, Rnt20, Robin Kerrison, Robma, Rory096, RoyBoy, RxS, Ryan Postlethwaite, Ryt, Ryulong, Ryūkotsusei, SDC, SMC, Saam 1, Sacasticnousisawesome, Saiswa, Salad Days, Salix alba, Salvio giuliano, Sam Hocevar, Sam Korn, Sander123, Sardanaphalus, Saruha, Savant13, Scarian, Sceptre, Schizobullet, Schutz, Scooter11, Scott14, Scriberius, Scrumshus, Sct72, Sdog 12, Sealpoint33, Sean D Martin, Seanwhim, Seb az86556, Selfworm, Servus Triviae, Sforzanda, Shabbat, Shadowjams, Shalom Yechiel, Shanel, Shaunwhim, Shii, Shoeofdeath, Shoessss, Shoy, Siener, Sietse Snel, Silly rabbit, Silverxxx, Simetrical, Singinglemon, SiobhanHansa, Sirtrebuchet, Sjakkalle, Sk238, Skant, Skarioffszky, Skateboy73, SkerHawx, Slakr, Sleeping123, SlightlyMad, Sligocki, Snowolf, Sockettome, Solipsist, Some P. Erson, Someguy1221, Sompm, Sopoforic, Sorrow* *sister of Joy, SpeedyGonsales, SpuriousQ, Squall-leon8, SquishedLizard, Sshadow, Stan Shebs, Stephen Sedgwick, Stephenb, Stephenchou0722, Stephensamuel, Steven Weston, Stevensaylor, Stevertigo, Student7, Stupid666, SummonerMarc, Superbeecat, Supersonic111, Suslindisambiguator, Sverdrup, Symane, Sześćsetsześćdziesiątsześć, Szlam, TOertel, Tachyon01, Tangotango, Tanvir Ahmmed, Tapir Terrific, Tarten5, Tbhotch, Tchoutoye, Tcnuk, Tdadamemd, Template namespace initialisation script, Terez27, Tex, The Epopt, The Man in Question, The Rambling Man, The Sage of Stamford, The enemies of god, The undertow, TheDaftman, TheDevil, TheMadBaron, Thedjatclubrock, Thehelpfulone, Thejerm, Themanganator7, Themusicking, Theone00, Theos theory, Thoscsii, Tiddly Tom, Tiger13, Timbouctou, Tjss, Tkuvho, Tlim7882, Tobias Hoevekamp, Tom Lougheed, TomPhil, Tomruen, TonyW, Toolazy21, Travelbird, Tree Biting Conspiracy, Tristessa de St Ange, Triwbe, Troels Nybo, Tv316, Twas Now, Twaz, Tzetzes, Ubardak, Uhfolks, Unfree, Ute in DC, Varanwal, Vaynor, Vcelloho, Vector4F, Vegan4peace, Venu62, Vikaszt, Viridae, VirtualDelight, Vitass13, Vittuone, Viznut, Vrenator, WIN, Waffles&n00dles, Wafulz, Waggers, Wahoofive, Wareh, Wassermann, Wavemaster447, Werdan7, Weregerbil, Westlysims, Whiskey in the Jar, Widefox, Wiki alf, WikiDao, WikiMichel, Wikipedia6469, William Avery, Wilt, Wimt, WinGeek, Wine Guy, Wintz75, Wolfgang975, Woohookitty, Wsiegmund, Wvurocks82, XJamRastafire, Yamamoto Ichiro, YellowMonkey, Yorkshirian, Yousou, Yummifruitbat, YuriLandman, Zae, Zhou yi777, Zippanova, Zundark, Zzuuzz, Александъp, , පසිඳු කාවින්ද, 2558 anonymous edits

Thales *Source*: http://en.wikipedia.org/w/index.php?title=Thales *Contributors*: .marc., 15lsoucy, 1984, 3rdAlcove, 65.197.2.xxx, 96T, A little insignificant, ABF, Acroterion, Adam Bishop, Adam Conover, Addshore, Aervanath, Aftermath, AgentPeppermint, Aitias, Alansohn, Alcmaeonid, Alcove, Aldux, AleksandrGertsen, Alex.muller, Allens, An Siarach, Andre Engels, Andrew Levine, Andrewkim99, Angielaj, Animum, Anna Lincoln, Anomie, Ashot Gabrielyan, At-par, Athenean, Atif.t2, Az1568, BD2412, Baristarim, Barticus88, Basketball110, Beavis5000, Belovedfreak, Bender235, Big iron, Biglovinb, Bkwillwm, BlindEagle, Bluemoose, Bobo192, Brion VIBBER, BrokenSegue, Bryan Derksen, Bubba73, Bullzeye, CHITRANI, Calmer Waters, Caltas, Can't sleep, clown will eat me, CanadianLinuxUser, Catalographer, Cedricbear, Chairboy, ChanDMan2010, Charles Matthews, Chetvorno, ChicXulub, Chochopk, Choess, Chowbok, Chrislk02, Chronicler, Chuunen Baka, Cje, ClockworkLunch, Codetiger, Cometstyles, CommonsDelinker, Conversion script, Cow Taurog, Cwoyte, D. Webb, DARTH SIDIOUS 2, DBigXray, DBodor, DMacks, DanielCD, Danski14, Dave6, Davidiad, Davitenio, Dawn Bard, Dcljr, Dedoleo, Deflective, DerHexer, Deville, Diderot, Dkastner, Dkderick, DocWatson42, Dommccas, Dougweller, Dreadstar, Dspradau, EPM, ESkog, EagleFan, Edderso, Eeekster, Eleuther, Elizabeyth, Epbr123, Epolk, Eric Kvaalen, Essent, Excirial, FallenAngel7789, FeanorStar7, Filippo Argenti, Finemann, First Light, Fleminra, FranksValli, Fryd1e, Fuhghettaboutit, Funnybunny, Fushigi-kun, GT5162, Gabloo, Gaelen S., Gaius Cornelius, Galanskov, Gat, Gatoatigrado, GcSwRhIc, Gchambonthales, Giftlite, Gilliam, Gja34, Glacialfox, Gpvos, Grafen, Greco22, Greenmangowango, GreyCat, Gscshoyru, Gummieworm, Gurch, Gwen-chan, Gwernol, Happy-melon, HarlandQPitt, Harry, Headbomb, Hemlock Martinis, Heracles31, HexaChord, Historybuff90, Hockeyrockcars, Hongooi, Hotshot977, Hu12, HunterX, Hyst, Icairns, Igiffin, Igoldste, ImperatorExercitus, Incnis Mrsi, Infosocialist, Insulanus, Intrr, Inwind, Iridescent, IronGargoyle, IslandHopper973, Isomorphic, Italus, Ixfd64, J.delanoy, JEN9841, JYolkowski, Jaellee, Jagged 85, Jaraalbe, Jaredwf, Jazgirl13, Jeandré du Toit, JesterCountess, Jim.henderson, JimWae, Jimp, Joannachan, JoanneB, Joevandy, John.d.page, JohnCD, Johnelwayrules, Johnwrw, Joseph Dwayne, Joseph Myers, Jossi, Jsf8204, Jsnx, Juc123, KRBN, Kalogeropoulos, Kazvorpal, Kevmanetti, Khazar, Killiondude, Kimon, Kingpin13, Knucmo2, Korg, Koxinga, Koyaanis Qatsi, Kristaga, Kungfuadam, Laureos, LeaveSleaves, Leoadec, Les boys, Lestrade, Levi allemany, Limexx35, Lir, Little Mountain 5, Llywrch, Logan, Lottamiata, LoveMonkey, Lowellian, Lrocky223, Lucidish, Luk, Lumos3, MDfoo, MER-C, Macedonian, Magioladitis, Magister Mathematicae, Mani1, Mariehuynh, Mark Krueger, Markcob, Martinmuse, Marudubshinki, Master of Puppets, Matusz, Maurice Carbonaro, Mav, Mccartjt, Measure for Measure, Melesse, Mendaliv, Mets501, Mhking, Miaow Miaow, Michael Hardy, MichaelTinkler, Michaelmas1957, Mig77, MisfitToys, Misza13, Mitartep, Moeron, Momo san, Moreschi, Mrdice, Mrjake, Mtpuls1092, Munita Prasad, N419BH, NATALEE--xD, Nareek, Naturyl, Ndaco, Nedim Ardoğa, Neoglyph, NewEnglandYankee, Nickbyrd, Nicke Lilltroll, Nigholith, Nixdorf, Nk, No such user, Noah Salzman, Noisy, Noname, Noqoilpi, NorthernDrifter, Novangelis, Ntsimp, NuclearWarfare, Nwwaew, Odysses, Oleg Alexandrov, Olivier, Omarraii, Oracleofottawa, Oxymoron83, PBS, PL290, Paine Ellsworth, Patricius Augustus, Paul August, Paulasiri2, Pbroks13, Pergamino, Peter cohen,

Peterbruce01, Peterlewis, Peyre, Pezanos, Pfalstad, Pfhenshaw, Philip Trueman, Piano non troppo, Pianodwarf, Piniricc65, Pollinosisss, Princess Lirin, Professor marginalia, Prolog, Puchiko, Qxz, R'n'B, RA0808, Radgeek, Reediewes, Reinis, Rettetast, RexNL, Rich Farmbrough, RichMac, Rickmorcom, RideLightning, Robin Hood, Roy Brumback, Rror, Rursus, Ryūkotsusei, S, Sango123, Scarian, Sceptre, Scimitar, Sciurinæ, Seans Potato Business, Secretlondon, Selfworm, Shadow1, Shadowjams, Shafei, Shanes, Shanoman, Silentlight, Simonides, Singinglemon, Smalljim, SnowFire, Some jerk on the Internet, Someguy1221, Sshadow, Starwiz, Stephenb, Stevenmitchell, Stockback, Struway, Sumahoy, Sustainablefutures2015, Tangent747, Tdw1203, Teddyw73, Tedmund, Template namespace initialisation script, Tevildo, Tgeairn, The Man in Question, The Sage of Stamford, The Thing That Should Not Be, Thesecondworldwar.net, This Is M4dn355 300, Thue, Thunderboltz, Tide rolls, Tpbradbury, Triwbe, Trumpetcase, Ugur Basak, Ukexpat, Velella, Versus22, VirtualDelight, VolatileChemical, Vrenator, WFinch, Waacstats, Wahabijaz, Waliddamouny, Welshwatch, Why Not A Duck, Wikiklrsc, Wikipelli, Wimt, Wknight94, Writtenright, XJamRastafire, Yamamoto Ichiro, Yiuchung, Zerida, 937 anonymous edits

Hypatia *Source*: http://en.wikipedia.org/w/index.php?title=Hypatia *Contributors*: 345Kai, 3rdAlcove, 545lljkr, 8digits, AWR Egypt, Aasmith, Ace of Spades, Adam Bishop, Adoarns, Ahoerstemeier, Ainlina, Aitias, Alf.laylah.wa.laylah, AliveFreeHappy, Allybobally, Amcfadgen, AnOddName, Andre Engels, Andreas Kaganov, Andrzej Kmicic, Angusmclellan, Antandrus, Anthony on Stilts, Antioco79, Anville, Arch4ngel, Arun9345, Athenean, Athkalani, Atkinson 291, Auranor, Average Earthman, Awwiki, BACbKA, BK 10000, BMBTHC, Baal wa astarte, Baroque Trumpet, Barroncd, Basicdesign, BehnamFarid, Benjamin Mako Hill, Bfigura, Big Brother 1984, Bill Thayer, Bloodofox, Bob Jonkman, Bobo192, Boffob, Bondolo, Bongwarrior, Bookworm857158367, Bookwormlm, Bped1985, BrainyBabe, Brandmeister, Bredwyn, Bridesmill, BridieO, Bwyche, CAWUEgypt, Cajunphysics, Caltas, Calu2000, Calvin 1998, Canadianism, Carl.bunderson, Carlo.Ierna, CatJar, Catalographer, Cerealkiller13, Chaojoker, Charles Matthews, Chnv, Cholchester1221, Christopher Mahan, CiaPan, Ckatz, Constcon, Courcelles, Cplakidas, Cuchullain, CuteHappyBrute, Cynwolfe, DHooke1973, DagosNavy, Dalf, Danger, Danny, Danski14, Darius Arcturus, David Eppstein, Davidiad, Dawn Bard, Dawnseeker2000, Dcoetzee, Deflective, Deor, Deutschgirl, Devil's Advocate, Dewrad, DianaGaleM, Dimadick, Dirkbb, Discospinster, Dj Capricorn, Djordjes, Dodgson, Donfbreed, Donreed, Dpr, Dr. Submillimeter, DrKamaila, Drexx, Drhoz, Druworos, Dtremenak, Dumaka, Duncharris, Duplode, Dwayne90210, ENeville, Eastlaw, EauLibrarian, Elassint, ElentariAchaea, ElinorD, Ellywa, Eloquence, Enigmaman, Epbr123, Erufailon42, Eskandarany, Eve Teschlemacher, Evercat, Ewulp, Favonian, FeanorStar7, Flamarande, Floaterfluss, Francescogiusti1, Frankie1969, FranksValli, Func, Future Perfect at Sunrise, Galo1969X, Gekritzl, GestaltG, Giftlite, Gilliam, Giovanni33, Goustien, GreatWhiteNortherner, Griot-de, Grizzledbear, Gspy, Hagerman, Hajor, Haluksomer, Hcbbowhuntr96, Heian-794, Henning Makholm, Heron, Hikingkyle393, Honbicot, Horologium, Hypathos, IXavier, Iain.dalton, Ian.thomson, Icairns, Igiffin, Ignacio Bibcraft, Inwind, Iothiania, Iphthime, IvanLanin, J.delanoy, J8079s, JForget, Ja 62, Jacob1207, Jagged 85, Jair Moreno, Jalaya101, Jallan, JamesBWatson, Jamesontai, Jaredwf, Jauhienij, Jaxsonjo, Jebus989, Jeudej, Jhaldenwang, Jic, JiggeryPokery, Jim.henderson, Jmlk17, John254, JohnChrysostom, Johnbibby, Jonathan Stokes, JorgeGG, Jtackney, Jumbuck, Justin Z, KConWiki, Kalidasa 777, Kansoku, Kazuba, Kbdank71, Kdavidk, Kerrym, Kevlarmry, Kikichef, Killercrossover, Kkm010, Klutzrick, Koavf, Kodster, Krisorey, Kwamikagami, Kyaku84, Lambiam, LauraMcKelvie, LaurenKaplow, Le Anh-Huy, Leinad-Z, Leonbravo, Lepidoptera, Lesybil, Lijealso, Lir, Lorynote, Lotje, LoveMonkey, Lucidish, Lucky number 49, Macedonian, MadGuy7023, Madeleineleigh, Madhero88, Maestlin, Magdalenemariefrylxky, Manytexts, Martin451, Materialscientist, Mathfriend, Matthew Fennell, Matthew Treder, Maurobio, Mcewan, Meeso, Megaleech, MelbourneStar, Menchi, Mercury McKinnon, Mhazard9, Michael Snow, MichaelTinkler, Michaelas10, Micione, Miguello80, Mike Rosoft, Mimihitam, MithrasPriest, Mladifilozof, Moe Epsilon, MoggieBleu, Momomamamesh, Monsieur Voltaire, Moretz, Morgan Leigh, Mu5ti, Murtasa, Musical Linguist, Myanw, Myasuda, Mystic.Ventus, N0osphR, N2e, N419BH, Namisada, Natalie Erin, NawlinWiki, Nevwik, NewEnglandYankee, Niceguyedc, Noalid, Notreallydavid, NovaDog, Nunh-huh, Oblivious, Ocaasi, Odysses, Ohyeawakeup, Oldlaptop321, Olivierdufault, Omnipaedista, Oskar71, Panarjedde, Panddpictures, Paolo Belzoni, Papamitsos, Parkjunwung, PatGallacher, Paul August, Paul Barlow, PaulLev, Peaceingalaxy, Pejman47, Peopledowhattheyoughttodo, Peroves, Peruvianllama, Peter.C, PeterStJohn, Petropoxy (Lithoderm Proxy), Petrouchka, Phe, Phil Boswell, Pigman, Pinaf, Pinethicket, PleaseStand, Plrk, Pohick2, Pollinosisss, Possum, Practical321, ProGloriaDei, Professor marginalia, Prosfilaes, Prsephone1674, Pshent, Punditah, Qrsdogg, Quantumobserver, Qwertyus, RB1956, RG2, Ragestorm, Ram Jaane, Ramyhamdy, Razimantv, Reach Out to the Truth, RedWolf, Redheylin, Richard Arthur Norton (1958-), Rick Norwood, Ricktерp, Ricky81682, Rjd0060, Rjwilmsi, Roadrunner, Roberpl, Robert1000, Robo89G, Roger Pearse, Rogper, RoyBoy, SJP, SU Linguist, Sam Korn, Samatva, Sander123, Sango123, SarahStierch, SchfiftyThree, Selfworm, ShakingSpirit, Shanalk, Shastrokes, Sherurcij, Shii, Shoeofdeath, Shsilver, Shtove, Singinglemon, Skarioffszky, Smjg, Soliderofsin, Soliloquial, Sonjaaa, Sp, Speciate, Startstop123, StaticVision, Stefanomione, Stellabystarlight, Stevebritgimp, Sthenel, Str1977, Strabismus, StradivariusTV, Suffusion of Yellow, Superbeecat, SuzanneIAM, Switchercat, Syncategoremata, TX55, Taiwan boi, Technopat, Tedmek, Thanatos666, TharkunColl, The High Fin Sperm Whale, The Thing That Should Not Be, TheNusz, Thecheesykid, Theo Pardilla, Thesisbinder, Thingg, Thorwald, Tide rolls, Titodutta, Tjepsen, Tom Peters, Tomisti, Tony Sidaway, Triune, Tvoz, Tyciol, Tzetzes, Uncle Dick, VI, Vanished user 2345, VasilievVV, Vegaswikian, VeryVerily, Viriditas, VivaEmilyDavies, WACourson, WLRoss, WP Editor 2011, Wavelength, Weetoddid, Wesley, Wetman, Whereizben, Wickedjacob, Wiki alf, Wiki13, WillowW, Willowj, Wilmot1, XAtalantax, XJamRastafire, Xenovatis, Xinstalker, Yamara, Yekrats, Yone Fernandes, Yooo12345, Yoooo12345, ZAROVE, ZackTheJack, Zidel333, Zidezi, Zorakoid, Þjóðólfr, , 824 anonymous edits

Geminus *Source*: http://en.wikipedia.org/w/index.php?title=Geminus *Contributors*: Amalas, Bungle, Catalographer, Ckatz, Conone, Damicatz, Deucalionite, Dingus1234, Dr. Submillimeter, Dysepsion, Ego White Tray, Felixboy, Filnik, Glane23, Hmains, Infoterra, JPfreak, Jaberwocky6669, Jaraalbe, Jgtl2, KRBN, Lupin, Maurog, Mike s, Mufka, Peruvianllama, Pezanos, Philip Trueman, RJHall, Rory096, Search4Lancer, Selfworm, Singinglemon, Skeptic2, That Guy, From That Show!, Tomisti, Tregonsee, Vamhendriks, XJamRastafire, 39 anonymous edits

Hippocrates_of_Chios *Source*: http://en.wikipedia.org/w/index.php?title=Hippocrates_of_Chios *Contributors*: Andycjp, Benwildeboer, Catalographer, Charles Matthews, ChazYork, Delirium, Djordjes, Gcm, Griot-de, Guitarskills, Gwern, Hans van Deukeren, Happy138, HendrixEesti, Ian Pitchford, JCSantos, Jaraalbe, Jbloomfield, Jdcooper, Lapaz, Macedonian, Magister Mathematicae, Marudubshinki, Merrill, Michael Hardy, Nettings, Nic bor, Nick C, Oracleofottawa, Pathoschild, Peruvianllama, Qmwne235, Ritchy, RockMFR, Sam Blacketer, Search4Lancer, Selfworm, Shivi1232423152, Singinglemon, Valentinian, Vary, XJamRastafire, Zad68, 51 anonymous edits

Image Sources, Licenses and Contributors

File:Aphrodite of Milos.jpg *Source*: http://en.wikipedia.org/w/index.php?title=File:Aphrodite_of_Milos.jpg *License*: unknown *Contributors*: User:Shawnlipowski

File:Greek Colonization.png *Source*: http://en.wikipedia.org/w/index.php?title=File:Greek_Colonization.png *License*: unknown *Contributors*: Based on a map in The Times History Atlas of the World

Image:Filip II Macedonia.jpg *Source*: http://en.wikipedia.org/w/index.php?title=File:Filip_II_Macedonia.jpg *License*: unknown *Contributors*: Gunnar Bach Pedersen

Image:BattleofIssus333BC-mosaic-detail1.jpg *Source*: http://en.wikipedia.org/w/index.php?title=File:BattleofIssus333BC-mosaic-detail1.jpg *License*: unknown *Contributors*: user:Ruthven

File:MacedonEmpire.jpg *Source*: http://en.wikipedia.org/w/index.php?title=File:MacedonEmpire.jpg *License*: unknown *Contributors*: Bjankuloski06en, Captain Blood, Doctor Che, Emijrp, Herbythyme, Jackaranga, Kallerna, LX, Pici3k, Shakko, SilvioRusso, Zykasaa, 17 anonymous edits

Image:Diadochen1.png *Source*: http://en.wikipedia.org/w/index.php?title=File:Diadochen1.png *License*: unknown *Contributors*: Captain_Blood

Image:Ptolemy I Soter Louvre Ma849.jpg *Source*: http://en.wikipedia.org/w/index.php?title=File:Ptolemy_I_Soter_Louvre_Ma849.jpg *License*: unknown *Contributors*: User:Jastrow

Image:Seleuco I Nicatore.JPG *Source*: http://en.wikipedia.org/w/index.php?title=File:Seleuco_I_Nicatore.JPG *License*: unknown *Contributors*: User:Finizio

Image:Seleucid-empire-323BCE.png *Source*: http://en.wikipedia.org/w/index.php?title=File:Seleucid-empire-323BCE.png *License*: unknown *Contributors*: self, based on WP locator maps

Image:PtolemaicEmpire.png *Source*: http://en.wikipedia.org/w/index.php?title=File:PtolemaicEmpire.png *License*: unknown *Contributors*: JMCC1, Neithsabes, World Imaging, Zykasaa

File:Smuseum 011hellenistic greek tomb door.jpg *Source*: http://en.wikipedia.org/w/index.php?title=File:Smuseum_011hellenistic_greek_tomb_door.jpg *License*: unknown *Contributors*: User:Storye book

File:Asia Minor 188 BCE.jpg *Source*: http://en.wikipedia.org/w/index.php?title=File:Asia_Minor_188_BCE.jpg *License*: unknown *Contributors*: Original uploader was Fishal at en.wikipedia

Image:Antiochos III.jpg *Source*: http://en.wikipedia.org/w/index.php?title=File:Antiochos_III.jpg *License*: unknown *Contributors*: Auguste Giraudon

Image:Greco-BactrianKingdomMap.jpg *Source*: http://en.wikipedia.org/w/index.php?title=File:Greco-BactrianKingdomMap.jpg *License*: unknown *Contributors*: Juiced lemon, World Imaging, Zaccarias, 1 anonymous edits

Image:Demetrius I MET coin.jpg *Source*: http://en.wikipedia.org/w/index.php?title=File:Demetrius_I_MET_coin.jpg *License*: unknown *Contributors*: User:Uploadalt

File:IndoGreekCampaings.jpg *Source*: http://en.wikipedia.org/w/index.php?title=File:IndoGreekCampaings.jpg *License*: unknown *Contributors*: User:PHGCOM

File:PonticKingdom.png *Source*: http://en.wikipedia.org/w/index.php?title=File:PonticKingdom.png *License*: unknown *Contributors*: Javierfv1212 (talk). Original uploader was Javierfv1212 at en.wikipedia

File:ancientlibraryalex.jpg *Source*: http://en.wikipedia.org/w/index.php?title=File:Ancientlibraryalex.jpg *License*: unknown *Contributors*: AnRo0002, Danmichaelo, Domitori, Gryffindor, JMCC1, Jakednb, Jmabel, Misa123a, Mmcannis, Olybrius, Quibik, Soerfm

File:PHAROS2006.jpg *Source*: http://en.wikipedia.org/w/index.php?title=File:PHAROS2006.jpg *License*: unknown *Contributors*: Emad Victor SHENOUDA (original uploader to de:)

Image:Colosse de Rhodes (Barclay).jpg *Source*: http://en.wikipedia.org/w/index.php?title=File:Colosse_de_Rhodes_(Barclay).jpg *License*: unknown *Contributors*: gravure sur bois de Sidney Barclay numérisée Google

File:Rosetta Stone.JPG *Source*: http://en.wikipedia.org/w/index.php?title=File:Rosetta_Stone.JPG *License*: unknown *Contributors*: User:Biopics

File:Temple Apollo Korinth OLC.jpg *Source*: http://en.wikipedia.org/w/index.php?title=File:Temple_Apollo_Korinth_OLC.jpg *License*: unknown *Contributors*: User:Olecorre

File:Zeno of Citium pushkin.jpg *Source*: http://en.wikipedia.org/w/index.php?title=File:Zeno_of_Citium_pushkin.jpg *License*: unknown *Contributors*: user:shakko

Image:Domenico-Fetti Archimedes 1620.jpg *Source*: http://en.wikipedia.org/w/index.php?title=File:Domenico-Fetti_Archimedes_1620.jpg *License*: unknown *Contributors*: A. Wagner, Andreagrossmann, AndreasPraefcke, Boo-Boo Baroo, Bukk, Christophe.Finot, FranzK, Gene.arboit, Ianmacm, Karel K., Kilom691, Kramer Associates, Luestling, Mattes, Plindenbaum, Serge Lachinov, Shakko, Wst, 3 anonymous edits

Image:Heron.jpeg *Source*: http://en.wikipedia.org/w/index.php?title=File:Heron.jpeg *License*: unknown *Contributors*: Delirium, Frank C. Müller, Heron

File:Laocoön and His Sons.jpg *Source*: http://en.wikipedia.org/w/index.php?title=File:Laocoön_and_His_Sons.jpg *License*: unknown *Contributors*: User:JuanMa

File:Domenico-Fetti Archimedes 1620.jpg *Source*: http://en.wikipedia.org/w/index.php?title=File:Domenico-Fetti_Archimedes_1620.jpg *License*: unknown *Contributors*: A. Wagner, Andreagrossmann, AndreasPraefcke, Boo-Boo Baroo, Bukk, Christophe.Finot, FranzK, Gene.arboit, Ianmacm, Karel K., Kilom691, Kramer Associates, Luestling, Mattes, Plindenbaum, Serge Lachinov, Shakko, Wst, 3 anonymous edits

File:Leonhard Euler.jpg *Source*: http://en.wikipedia.org/w/index.php?title=File:Leonhard_Euler.jpg *License*: unknown *Contributors*: User:Wars

File:Carl Friedrich Gauss.jpg *Source*: http://en.wikipedia.org/w/index.php?title=File:Carl_Friedrich_Gauss.jpg *License*: unknown *Contributors*: Gottlieb BiermannA. Wittmann (photo)

File:Hilbert.jpg *Source*: http://en.wikipedia.org/w/index.php?title=File:Hilbert.jpg *License*: unknown *Contributors*: DaTroll, Darapti, Der Eberswalder, Gene.arboit, Harp, Mschlindwein, Yann, Zwikki, 3 anonymous edits

File:Noether.jpg *Source*: http://en.wikipedia.org/w/index.php?title=File:Noether.jpg *License*: unknown *Contributors*: Anarkman, Awadewit, Darapti, Factumquintus, Gene.arboit, Lambiam, Lobo, Materialscientist, Romary

File:GodfreyKneller-IsaacNewton-1689.jpg *Source*: http://en.wikipedia.org/w/index.php?title=File:GodfreyKneller-IsaacNewton-1689.jpg *License*: unknown *Contributors*: Algorithme, Beyond My Ken, Bjankuloski06en, Grenavitar, Infrogmation, Kelson, Kilom691, Porao, Saperaud, Semnoz, Siebrand, Sparkit, Thomas Gun, Vonvon, Wikiklaas, Wknight94, Wst, Zaphod, 7 anonymous edits

File:Bust Hadrian Musei Capitolini MC817.jpg *Source*: http://en.wikipedia.org/w/index.php?title=File:Bust_Hadrian_Musei_Capitolini_MC817.jpg *License*: unknown *Contributors*: User:Jastrow

File:Roman Empire 125.png *Source*: http://en.wikipedia.org/w/index.php?title=File:Roman_Empire_125.png *License*: unknown *Contributors*: User:Andrein, User:EraNavigator

File:Hadrians Wall from Housesteads1.jpg *Source*: http://en.wikipedia.org/w/index.php?title=File:Hadrians_Wall_from_Housesteads1.jpg *License*: unknown *Contributors*: Original uploader was Jamesflomonosoff at en.wikipedia

File:Hadrianus gate.jpg *Source*: http://en.wikipedia.org/w/index.php?title=File:Hadrianus_gate.jpg *License*: unknown *Contributors*: Alexandrin, AnRo0002, FlickreviewR, Freta, G.dallorto, Nosferatü, Sunrise1, 2 anonymous edits

File:Athenstemplezeus.jpg *Source*: http://en.wikipedia.org/w/index.php?title=File:Athenstemplezeus.jpg *License*: unknown *Contributors*: Bibi Saint-Pol, Helentr, Jeanhousen, Laurascudder, Madmedea

File:Pantheon Agrippa.jpg *Source*: http://en.wikipedia.org/w/index.php?title=File:Pantheon_Agrippa.jpg *License*: unknown *Contributors*: Briséis, Sailko, 1 anonymous edits

File:Adriano5.jpg *Source*: http://en.wikipedia.org/w/index.php?title=File:Adriano5.jpg *License*: unknown *Contributors*: User:Tetraktys

File:Conic sections with plane.svg *Source*: http://en.wikipedia.org/w/index.php?title=File:Conic_sections_with_plane.svg *License*: unknown *Contributors*: User:Pbroks13

File:Table of Conics, Cyclopaedia, volume 1, p 304, 1728.jpg *Source*: http://en.wikipedia.org/w/index.php?title=File:Table_of_Conics,_Cyclopaedia,_volume_1,_p_304,_1728.jpg *License*: unknown *Contributors*: Brian0918, LaosLos, Tano4595

File:Conic Sections.svg *Source*: http://en.wikipedia.org/w/index.php?title=File:Conic_Sections.svg *License*: unknown *Contributors*: Phancy Physicist

File:Ellipse parameters 2.svg *Source*: http://en.wikipedia.org/w/index.php?title=File:Ellipse_parameters_2.svg *License*: unknown *Contributors*: User:Bomazi

File:Conic sections3.svg *Source*: http://en.wikipedia.org/w/index.php?title=File:Conic_sections3.svg *License*: unknown *Contributors*: zinka

File:Archeocyathids.JPG *Source*: http://en.wikipedia.org/w/index.php?title=File:Archeocyathids.JPG *License*: unknown *Contributors*: User:Citypeek

File:Parabola connection with areas of a square and a rectangle.gif *Source*: http://en.wikipedia.org/w/index.php?title=File:Parabola_connection_with_areas_of_a_square_and_a_rectangle.gif *License*: unknown *Contributors*: User:Cmapm

File:Gerhard Thieme Archimedes.jpg *Source*: http://en.wikipedia.org/w/index.php?title=File:Gerhard_Thieme_Archimedes.jpg *License*: unknown *Contributors*: Editor at Large, Gerardus, Ianmacm, Kilom691, Lobo, Lotse, SpreeTom, Srittau, Wst, 3 anonymous edits

File:Archimedes-screw one-screw-threads with-ball 3D-view animated small.gif *Source*: http://en.wikipedia.org/w/index.php?title=File:Archimedes-screw_one-screw-threads_with-ball_3D-view_animated_small.gif *License*: unknown *Contributors*: User:Jahobr, User:Silberwolf

File:Archimedes Heat Ray conceptual diagram.svg *Source*: http://en.wikipedia.org/w/index.php?title=File:Archimedes_Heat_Ray_conceptual_diagram.svg *License*: unknown *Contributors*: User:Finnrind, User:Pbroks13

File:PiArchimede4.svg *Source*: http://en.wikipedia.org/w/index.php?title=File:PiArchimede4.svg *License*: unknown *Contributors*: User:HB

File:Stomachion.JPG *Source*: http://en.wikipedia.org/w/index.php?title=File:Stomachion.JPG *License*: unknown *Contributors*: Ianmacm at en.wikipedia

Image:Diophantus-cover.jpg *Source*: http://en.wikipedia.org/w/index.php?title=File:Diophantus-cover.jpg *License*: unknown *Contributors*: Khadkhall, Lord Akryl, Mdd, Mu, Proteins, Schutz, 1 anonymous edits

File:Scuola di atene 07.jpg *Source*: http://en.wikipedia.org/w/index.php?title=File:Scuola_di_atene_07.jpg *License*: unknown *Contributors*: Auntof6, Sailko

File:EuclidStatueOxford.jpg *Source*: http://en.wikipedia.org/w/index.php?title=File:EuclidStatueOxford.jpg *License*: unknown *Contributors*: User:Wilson44691

Image:wikisource-logo.svg *Source*: http://en.wikipedia.org/w/index.php?title=File:Wikisource-logo.svg *License*: unknown *Contributors*: Guillom, Jarekt, MichaelMaggs, NielsF, Rei-artur, Rocket000

File:Hipparchus by Raphael.jpg *Source*: http://en.wikipedia.org/w/index.php?title=File:Hipparchus_by_Raphael.jpg *License*: unknown *Contributors*: Raphael

File:HipparchusConstruction.png *Source*: http://en.wikipedia.org/w/index.php?title=File:HipparchusConstruction.png *License*: unknown *Contributors*: Foroa, JMCC1, Maksim, Roymail

File:HipparchusEclipse.png *Source*: http://en.wikipedia.org/w/index.php?title=File:HipparchusEclipse.png *License*: unknown *Contributors*: Maksim, 2 anonymous edits

File:Heron.jpeg *Source*: http://en.wikipedia.org/w/index.php?title=File:Heron.jpeg *License*: unknown *Contributors*: Delirium, Frank C. Müller, Heron

Image:Aeolipile_illustration.png *Source*: http://en.wikipedia.org/w/index.php?title=File:Aeolipile_illustration.png *License*: unknown *Contributors*: The entry under Aeolipile in volume one of this work states "The cut is copied from Hero's "Spiritalia", edited by Woodcroft, of London."

Image:heron3.jpg *Source*: http://en.wikipedia.org/w/index.php?title=File:Heron3.jpg *License*: unknown *Contributors*: Paxse, Peterlewis, 2 anonymous edits

File:Pythagoras Bust Vatican Museum.jpg *Source*: http://en.wikipedia.org/w/index.php?title=File:Pythagoras_Bust_Vatican_Museum.jpg *License*: unknown *Contributors*: Original uploader was Andargor at en.wikipedia

File:Pythagoras-Münz.JPG *Source*: http://en.wikipedia.org/w/index.php?title=File:Pythagoras-Münz.JPG *License*: unknown *Contributors*: Bibi Saint-Pol, Carlomorino, Schaengel89, 1 anonymous edits

File:Pythag anim.gif *Source*: http://en.wikipedia.org/w/index.php?title=File:Pythag_anim.gif *License*: unknown *Contributors*: User:JohnBlackburne

File:Sanzio 01 Pythagoras.jpg *Source*: http://en.wikipedia.org/w/index.php?title=File:Sanzio_01_Pythagoras.jpg *License*: unknown *Contributors*: Auntof6, Bibi Saint-Pol, Bukk, G.dallorto, Jacobolus, Kalki, Mattes, Pasicles, Sailko, Warburg, 1 anonymous edits

File:Nuremberg chronicles f 61v 2.png *Source*: http://en.wikipedia.org/w/index.php?title=File:Nuremberg_chronicles_f_61v_2.png *License*: unknown *Contributors*: Michel Wolgemut, Wilhelm Pleydenwurff (Text: Hartmann Schedel)

File:Thales.jpg *Source*: http://en.wikipedia.org/w/index.php?title=File:Thales.jpg *License*: unknown *Contributors*: Bibi Saint-Pol, Tomisti

Image:Capernaum roman olive press by David Shankbone.jpg *Source*: http://en.wikipedia.org/w/index.php?title=File:Capernaum_roman_olive_press_by_David_Shankbone.jpg *License*: unknown *Contributors*: User:DavidShankbone

Image:Solar eclipse 1999 4 NR.jpg *Source*: http://en.wikipedia.org/w/index.php?title=File:Solar_eclipse_1999_4_NR.jpg *License*: unknown *Contributors*: user:Lviatour

Image:Thales theorem 1.png *Source*: http://en.wikipedia.org/w/index.php?title=File:Thales_theorem_1.png *License*: unknown *Contributors*: Dake, Darapti, HB, Helder.wiki, Siebrand, 2 anonymous edits

Image:Louisstgaudens1.jpg *Source*: http://en.wikipedia.org/w/index.php?title=File:Louisstgaudens1.jpg *License*: unknown *Contributors*: Original uploader was Danvera at en.wikipedia

File:Hypatia (Charles William Mitchell).jpg *Source*: http://en.wikipedia.org/w/index.php?title=File:Hypatia_(Charles_William_Mitchell).jpg *License*: unknown *Contributors*: AnonMoos, Bibi Saint-Pol, Dodo, FlagUploader, Frank C. Müller, Gene.arboit, Infrogmation, Jarekt, Longhairadmirer, Mattes, Pasicles, Pilettes, Sammyday, Shakko, Sherurcij, Starscream, VladiMens, Zolo

File:Hipocrat arcs.svg *Source*: http://en.wikipedia.org/w/index.php?title=File:Hipocrat_arcs.svg *License*: unknown *Contributors*: Audriusa, Rimshot, Schneau, 1 anonymous edits

PREAMBLE

he purpose of this License is to make a manual, textbook, or ier functional and useful document "free" in the sense of edom: to assure everyone the effective freedom to copy and distribute it, with or without modifying it, either commercially or ncommercially. Secondarily, this License preserves for the thor and publisher a way to get credit for their work, while not ng considered responsible for modifications made by others. is License is a kind of "copyleft", which means that derivative rks of the document must themselves be free in the same nse. It complements the GNU General Public License, which is copyleft license designed for free software. We have designed s License in order to use it for manuals for free software, cause free software needs free documentation: a free program ould come with manuals providing the same freedoms that the tware does. But this License is not limited to software manuals; can be used for any textual work, regardless of subject matter whether it is published as a printed book. We recommend this ense principally for works whose purpose is instruction or erence.

APPLICABILITY AND DEFINITIONS

his License applies to any manual or other work, in any edium, that contains a notice placed by the copyright holder ying it can be distributed under the terms of this License. Such notice grants a world-wide, royalty-free license, unlimited in ration, to use that work under the conditions stated herein. The ocument", below, refers to any such manual or work. Any ember of the public is a licensee, and is addressed as "you". u accept the license if you copy, modify or distribute the work a way requiring permission under copyright law. A "Modified ersion" of the Document means any work containing the ocument or a portion of it, either copied verbatim, or with odifications and/or translated into another language. A econdary Section" is a named appendix or a front-matter ction of the Document that deals exclusively with the lationship of the publishers or authors of the Document to the ocument's overall subject (or to related matters) and contains thing that could fall directly within that overall subject. (Thus, if e Document is in part a textbook of mathematics, a Secondary ection may not explain any mathematics.) The relationship could a matter of historical connection with the subject or with lated matters, or of legal, commercial, philosophical, ethical or olitical position regarding them. The "Invariant Sections" are rtain Secondary Sections whose titles are designated, as being ose of Invariant Sections, in the notice that says that the ocument is released under this License. If a section does not fit e above definition of Secondary then it is not allowed to be esignated as Invariant. The Document may contain zero variant Sections. If the Document does not identify any Invariant ections then there are none. The "Cover Texts" are certain short assages of text that are listed, as Front-Cover Texts or Back-Cover Texts, in the notice that says that the Document is eleased under this License. A Front-Cover Text may be at most words, and a Back-Cover Text may be at most 25 words. A 'ransparent" copy of the Document means a machine-readable opy, represented in a format whose specification is available to e general public, that is suitable for revising the document traightforwardly with generic text editors or (for images omposed of pixels) generic paint programs or (for drawings) ome widely available drawing editor, and that is suitable for input text formatters or for automatic translation to a variety of rmats suitable for input to text formatters. A copy made in an therwise Transparent file format whose markup, or absence of arkup, has been arranged to thwart or discourage subsequent odification by readers is not Transparent. An image format is ot Transparent if used for any substantial amount of text. A copy nat is not "Transparent" is called "Opaque". Examples of suitable ormats for Transparent copies include plain ASCII without narkup, Texinfo input format, LaTeX input format, SGML or XML sing a publicly available DTD, and standard-conforming simple HTML, PostScript or PDF designed for human modification. Examples of transparent image formats include PNG, XCF and PG. Opaque formats include proprietary formats that can be ead and edited only by proprietary word processors, SGML or XML for which the DTD and/or processing tools are not generally available, and the machine-generated HTML, PostScript or PDF roduced by some word processors for output purposes only. The Title Page" means, for a printed book, the title page itself, plus such following pages as are needed to hold, legibly, the material his License requires to appear in the title page. For works in ormats which do not have any title page as such, "Title Page" neans the text near the most prominent appearance of the work's itle, preceding the beginning of the body of the text. A section 'Entitled XYZ" means a named subunit of the Document whose itle either is precisely XYZ or contains XYZ in parentheses ollowing text that translates XYZ in another language. (Here XYZ stands for a specific section name mentioned below, such as 'Acknowledgements", "Dedications", "Endorsements", or 'History".) To "Preserve the Title" of such a section when you nodify the Document means that it remains a section "Entitled XYZ" according to this definition. The Document may include Warranty Disclaimers next to the notice which states that this License applies to the Document. These Warranty Disclaimers are considered to be included by reference in this License, but only as regards disclaiming warranties: any other implication that these Warranty Disclaimers may have is void and has no effect on the meaning of this License.

2. VERBATIM COPYING

You may copy and distribute the Document in any medium, either commercially or noncommercially, provided that this License, the copyright notices, and the license notice saying this License applies to the Document are reproduced in all copies, and that you add no other conditions whatsoever to those of this License. You may not use technical measures to obstruct or control the reading or further copying of the copies you make or distribute. However, you may accept compensation in exchange for copies. If you distribute a large enough number of copies you must also follow the conditions in section 3. You may also lend copies, under the same conditions stated above, and you may publicly display copies.

3. COPYING IN QUANTITY

If you publish printed copies (or copies in media that commonly have printed covers) of the Document, numbering more than 100, and the Document's license notice requires Cover Texts, you must enclose the copies in covers that carry, clearly and legibly, all these Cover Texts: Front-Cover Texts on the front cover, and Back-Cover Texts on the back cover. Both covers must also clearly and legibly identify you as the publisher of these copies. The front cover must present the full title with all words of the title equally prominent and visible. You may add other material on the covers in addition. Copying with changes limited to the covers, as long as they preserve the title of the Document and satisfy these conditions, can be treated as verbatim copying in other respects. If the required texts for either cover are too voluminous to fit legibly, you should put the first ones listed (as many as fit reasonably) on the actual cover, and continue the rest onto adjacent pages. If you publish or distribute Opaque copies of the Document numbering more than 100, you must either include a machine-readable Transparent copy along with each Opaque copy, or state in or with each Opaque copy a computer-network location from which the general network-using public has access to download using public-standard network protocols a complete Transparent copy of the Document, free of added material. If you use the latter option, you must take reasonably prudent steps, when you begin distribution of Opaque copies in quantity, to ensure that this Transparent copy will remain thus accessible at the stated location until at least one year after the last time you distribute an Opaque copy (directly or through your agents or retailers) of that edition to the public. It is requested, but not required, that you contact the authors of the Document well before redistributing any large number of copies, to give them a chance to provide you with an updated version of the Document.

4. MODIFICATIONS

You may copy and distribute a Modified Version of the Document under the conditions of sections 2 and 3 above, provided that you release the Modified Version under precisely this License, with the Modified Version filling the role of the Document, thus licensing distribution and modification of the Modified Version to whoever possesses a copy of it. In addition, you must do these things in the Modified Version: A. Use in the Title Page (and on the covers, if any) a title distinct from that of the Document, and from those of previous versions (which should, if there were any, be listed in the History section of the Document). You may use the same title as a previous version if the original publisher of that version gives permission. B. List on the Title Page, as authors, one or more persons or entities responsible for authorship of the modifications in the Modified Version, together with at least five of the principal authors of the Document (all of its principal authors, if it has fewer than five), unless they release you from this requirement. C. State on the Title page the name of the publisher of the Modified Version, as the publisher. D. Preserve all the copyright notices of the Document. E. Add an appropriate copyright notice for your modifications adjacent to the other copyright notices. F. Include, immediately after the copyright notices, a license notice giving the public permission to use the Modified Version under the terms of this License, in the form shown in the Addendum below. G. Preserve in that license notice the full lists of Invariant Sections and required Cover Texts given in the Document's license notice. H. Include an unaltered copy of this License. I. Preserve the section Entitled "History", Preserve its Title, and add to it an item stating at least the title, year, new authors, and publisher of the Modified Version as given on the Title Page. If there is no section Entitled "History" in the Document, create one stating the title, year, authors, and publisher of the Document as given on its Title Page, then add an item describing the Modified Version as stated in the previous sentence. J. Preserve the network location, if any, given in the Document for public access to a Transparent copy of the Document, and likewise the network locations given in the Document for previous versions it was based on. These may be placed in the "History" section. You may omit a network location for a work that was published at least four years before the Document itself, or if the original publisher of the version it refers to gives permission. K. For any section Entitled "Acknowledgements" or "Dedications", Preserve the Title of the section, and preserve in the section all the substance and tone of each of the contributor acknowledgements and/or dedications given therein. L. Preserve all the Invariant Sections of the Document, unaltered in their text and in their titles. Section numbers or the equivalent are not considered part of the section titles. M. Delete any section Entitled "Endorsements". Such a section may not be included in the Modified Version. N. Do not retitle any existing section to be Entitled "Endorsements" or to conflict in title with any Invariant Section. O. Preserve any Warranty Disclaimers. If the Modified Version includes new front-matter sections or appendices that qualify as Secondary Sections and contain no material copied from the Document, you may at your option designate some or all of these sections as invariant. To do this, add their titles to the list of Invariant Sections in the Modified Version's license notice. These titles must be distinct from any other section titles. You may add a section Entitled "Endorsements", provided it contains nothing but endorsements of your Modified Version by various parties--for example, statements of peer review or that the text has been approved by an organization as the authoritative definition of a standard. You may add a passage of up to five words as a Front-Cover Text, and a passage of up to 25 words as a Back-Cover Text, to the end of the list of Cover Texts in the Modified Version. Only one passage of Front-Cover Text and one of Back-Cover Text may be added by (or through arrangements made by) any one entity. If the Document already includes a cover text for the same cover, previously added by you or by arrangement made by the same entity you are acting on behalf of, you may not add another; but you may replace the old one, on explicit permission from the previous publisher that added the old one. The author(s) and publisher(s) of the Document do not by this License give permission to use their names for publicity for or to assert or imply endorsement of any Modified Version.

5. COMBINING DOCUMENTS

You may combine the Document with other documents released under this License, under the terms defined in section 4 above for modified versions, provided that you include in the combination all of the Invariant Sections of all of the original documents, unmodified, and list them all as Invariant Sections of your combined work in its license notice, and that you preserve all their Warranty Disclaimers. The combined work need only contain one copy of this License, and multiple identical Invariant Sections may be replaced with a single copy. If there are multiple Invariant Sections with the same name but different contents, make the title of each such section unique by adding at the end of it, in parentheses, the name of the original author or publisher of that section if known, or else a unique number. Make the same adjustment to the section titles in the list of Invariant Sections in the license notice of the combined work. In the combination, you must combine any sections Entitled "History" in the various original documents, forming one section Entitled "History"; likewise combine any sections Entitled "Acknowledgements", and any sections Entitled "Dedications". You must delete all sections Entitled "Endorsements".

6. COLLECTIONS OF DOCUMENTS

You may make a collection consisting of the Document and other documents released under this License, and replace the individual copies of this License in the various documents with a single copy that is included in the collection, provided that you follow the rules of this License for verbatim copying of each of the documents in all other respects. You may extract a single document from such a collection, and distribute it individually under this License, provided you insert a copy of this License into the extracted document, and follow this License in all other respects regarding verbatim copying of that document.

7. AGGREGATION WITH INDEPENDENT WORKS

A compilation of the Document or its derivatives with other separate and independent documents or works, in or on a volume of a storage or distribution medium, is called an "aggregate" if the copyright resulting from the compilation is not used to limit the legal rights of the compilation's users beyond what the individual works permit. When the Document is included in an aggregate, this License does not apply to the other works in the aggregate which are not themselves derivative works of the Document. If the Cover Text requirement of section 3 is applicable to these copies of the Document, then if the Document is less than one half of the entire aggregate, the Document's Cover Texts may be placed on covers that bracket the Document within the aggregate, or the electronic equivalent of covers if the Document is in electronic form. Otherwise they must appear on printed covers that bracket the whole aggregate.

8. TRANSLATION

Translation is considered a kind of modification, so you may distribute translations of the Document under the terms of section 4. Replacing Invariant Sections with translations requires special permission from their copyright holders, but you may include translations of some or all Invariant Sections in addition to the original versions of these Invariant Sections. You may include a translation of this License, and all the license notices in the Document, and any Warranty Disclaimers, provided that you also include the original English version of this License and the original versions of those notices and disclaimers. In case of a disagreement between the translation and the original version of this License or a notice or disclaimer, the original version will prevail. If a section in the Document is Entitled "Acknowledgements", "Dedications", or "History", the requirement (section 4) to Preserve its Title (section 1) will typically require changing the actual title.

9. TERMINATION

You may not copy, modify, sublicense, or distribute the Document except as expressly provided for under this License. Any other attempt to copy, modify, sublicense or distribute the Document is void, and will automatically terminate your rights under this License. However, parties who have received copies, or rights, from you under this License will not have their licenses terminated so long as such parties remain in full compliance.

10. FUTURE REVISIONS OF THIS LICENSE

The Free Software Foundation may publish new, revised versions of the GNU Free Documentation License from time to time. Such new versions will be similar in spirit to the present version, but may differ in detail to address new problems or concerns. See http://www.gnu.org/copyleft/. Each version of the License is given a distinguishing version number. If the Document specifies that a particular numbered version of this License "or any later version" applies to it, you have the option of following the terms and conditions either of that specified version or of any later version that has been published (not as a draft) by the Free Software Foundation. If the Document does not specify a version number of this License, you may choose any version ever published (not as a draft) by the Free Software Foundation. ADDENDUM: How to use this License for your documents To use this License in a document you have written, include a copy of the License in the document and put the following copyright and license notices just after the title page: Copyright (c) YEAR YOUR NAME. Permission is granted to copy, distribute and/or modify this document under the terms of the GNU Free Documentation License, Version 1.2 or any later version published by the Free Software Foundation; with no Invariant Sections, no Front-Cover Texts, and no Back-Cover Texts. A copy of the license is included in the section entitled "GNU Free Documentation License". If you have Invariant Sections, Front-Cover Texts and Back-Cover Texts, replace the "with...Texts." line with this: with the Invariant Sections being LIST THEIR TITLES, with the Front-Cover Texts being LIST, and with the Back-Cover Texts being LIST. If you have Invariant Sections without Cover Texts, or some other combination of the three, merge those two alternatives to suit the situation. If your document contains nontrivial examples of program code, we recommend releasing these examples in parallel under your choice of free software license, such as the GNU General Public License, to permit their use in free software.

MIX
Papier aus verantwortungsvollen Quellen
Paper from responsible sources
FSC® C105338

Printed by Books on Demand GmbH, Norderstedt / Germany